Blockchain Potential in AI

*Edited by Tiago M. Fernández-Caramés
and Paula Fraga-Lamas*

Published in London, United Kingdom

IntechOpen

Supporting open minds since 2005

Blockchain Potential in AI
http://dx.doi.org/10.5772/intechopen.91580
Edited by Tiago M. Fernández-Caramés and Paula Fraga-Lamas

Contributors
Stanton Heister, Kristi Yuthas, Eva R. Porras, Bryan Daugherty, Hyun-joo Jeon, Ho-chang Youn, Sang-mi Ko, Tae-heon Kim, Zehua Wang, Victoria Lemieux, Yao Du, Shuxiao Miao, Zitian Tong, Tiago M. Fernández-Caramés, Paula Fraga-Lamas

Notice
Statements and opinions expressed in the chapters are these of the individual contributors and not necessarily those of the editors or publisher. No responsibility is accepted for the accuracy of information contained in the published chapters. The publisher assumes no responsibility for any damage or injury to persons or property arising out of the use of any materials, instructions, methods or ideas contained in the book.

First published in London, United Kingdom, 2022 by IntechOpen
IntechOpen is the global imprint of INTECHOPEN LIMITED, registered in England and Wales, registration number: 11086078, 5 Princes Gate Court, London, SW7 2QJ, United Kingdom
Printed in Croatia

British Library Cataloguing-in-Publication Data
A catalogue record for this book is available from the British Library

Additional hard and PDF copies can be obtained from orders@intechopen.com

Blockchain Potential in AI
Edited by Tiago M. Fernández-Caramés and Paula Fraga-Lamas
p. cm.
Print ISBN 978-1-78984-093-3
Online ISBN 978-1-78984-094-0
eBook (PDF) ISBN 978-1-78985-553-1

We are IntechOpen,
the world's leading publisher of
Open Access books
Built by scientists, for scientists

5,600+
Open access books available

138,000+
International authors and editors

175M+
Downloads

156
Countries delivered to

Our authors are among the
Top 1%
most cited scientists

12.2%
Contributors from top 500 universities

Interested in publishing with us?
Contact book.department@intechopen.com

Numbers displayed above are based on latest data collected.
For more information visit www.intechopen.com

Meet the editors

Tiago M. Fernández-Caramés (S'08-M'12-SM'15) works as an associate professor at the University of A Coruña (UDC), Spain, where he obtained his MSc degree and Ph.D. degrees in Computer Science. He has worked in the Department of Computer Engineering at UDC: from 2005 to 2009 through different predoctoral scholarships and, in parallel, since 2007, as a professor. His current research interests include blockchain, internet of things (IoT) and industrial IoT (IIoT) systems, wireless sensor networks, and augmented and mixed reality (AR/MR), as well as the different technologies involved in Industry 4.0 and Industry 5.0 paradigms. In such fields, he has contributed to more than a hundred papers for conferences and high-impact journals and books and has six patents.

Paula Fraga-Lamas (M'17, SM'20), Ph.D.-MBA in Computer Engineering, is a senior researcher and lecturer at the University of A Coruña (UDC), Spain. She has over a hundred contributions in indexed international journals, conferences, and book chapters, and three patents. She has participated in over 30 research projects funded by the regional and national government and has research and development (R&D) contracts with private companies. She is actively involved in many professional and editorial activities. She has worked as an expert evaluator and adviser of several national and international agencies. Her current research interests include Industry 5.0, internet of things (IoT), cyber-physical systems (CPS), augmented and mixed reality (AR/MR), edge computing, blockchain, and distributed ledger technologies (DLT), and cybersecurity.

Contents

Preface

Blockchain (BC) and artificial intelligence (AI) are currently two of the hottest computer science topics and their future seems bright. However, their convergence is not straightforward and more research is needed in both fields. Thus, this book presents some of the latest advances in the convergence of BC and AI, gives useful guidelines for future researchers on how BC can help AI and how AI can become smarter, thanks to the use of BC.

Specifically, Chapter 1 introduces the basics of the convergence of BC and AI and indicates the main opportunities and challenges of such a convergence.

Chapter 2 reviews the history of Bitcoin (the first blockchain-based cryptocurrency) and its influence on the market and on society. Thus, this chapter reviews the past with the objective of understanding how the convergence of BC and AI can impact our future, especially in terms of privacy and democracy.

Chapter 3 deals with how BC and AI can help in the cybersecurity field. In particular, the chapter explores how BC and AI can join forces to provide solutions to protect personal data. Such solutions should allow users to control how their personal information is accessed and to know who accessed such information. AI is essential for these solutions since it can complement BC-based applications when managing data and guaranteeing that the models obtained from such data are accurate, fair, and reliable.

Chapter 4 focuses on solving the challenges of resource-constrained internet-of-things (IoT) devices with mobile edge computing (MEC), which offloads part of the processing tasks from the cloud (e.g., complex learning tasks) to the edge. The chapter specifically proposes the creation of BC-enabled mobile edge intelligence in IoT scenarios. In addition, it also reviews the state of the art of the combination of BC and AI in different scenarios and analyzes the main security and privacy features of BC.

Furthermore, some of the threats to be faced by BC systems are analyzed. For example, the threat of quantum computing is considered, based on post-quantum blockchain solutions.

Metaverse, a shared virtual environment that combines augmented reality (AR) and virtual reality (VR), is introduced in Chapter 5 as an ever-expanding world where AI and BC will play a key role in the near future. The chapter describes the combination of the real and virtual worlds and the main challenges that may arise from such a combination (e.g., the reliability of the data). It proposes a BC-based metaverse using Ethereum that will enable safe and free social and economical activities and the secured application of technologies such as AI. In addition, the benefits that AI and BC can provide to the metaverse are also analyzed:

high-quality learning data, reusable data, stable decentralized network, privacy, the distinction between virtual and real information (including deepfakes and fake news), rich content, and reliable transactions and a trustworthy marketplace.

Tiago M. Fernández-Caramés and Paula Fraga-Lamas
Group of Electronic Technology and Communications (GTEC),
Faculty of Computer Science,
Universidade da Coruña,
A Coruña, Spain

Introductory Chapter: On the Convergence of Blockchain and Artificial Intelligence - Opportunities and Challenges

Paula Fraga-Lamas and Tiago M. Fernández-Caramés

1. Introduction

According to a PwC global study presented in 2019, Artificial Intelligence (AI) has the potential to contribute up to $ 15.7 trillion to the global economy by 2030 [1]. In a report of 2020, the same company predicted that blockchain (BC) applications will boost global Gross Domestic Product (GDP) in 2030 by $ 1.76 trillion (1.4% of global GDP) [2]. Such a boost in economic value in the next decade shows the potential of both technologies as key drivers of the current digital transformation.

AI, specifically the most used techniques today, Machine Learning (ML), Deep Learning (DL) or Reinforcement Learning (RL) can create learning models that process and analyze data, perform tasks or make predictions for diverse real-world problems that were previously thought to be impossible to be solved by nonhumans. However, the use of AI comes with social concerns related to issues like the possibility of data tampering due to data centralization, the rise of fake news and deep fakes [3], the invasion of privacy or the bias in data training.

Distributed Ledger Technology (DLTs) and specifically BC, can create trust and consensus among a group of participants removing the need for intermediaries. On the one hand, BC can help to establish the data provenance for explainable AI and to ensure the authenticity and reliability of the data sources used in techniques [4]. In addition, decentralized computing for AI enables decision-making on secured shared data in a decentralized manner without intermediaries. Furthermore, autonomous systems that make use of smart contracts can learn over time and make trusted decisions [5]. On the other hand, AI can help to face some of the current limitations in BC implementations like scalability or security, privacy-preserving personalization, automated refereeing, and governance mechanisms [6].

Both AI and BC are increasingly being used in similar and even in the same applications. Therefore, it is expected that AI and blockchain converge into BC-AI systems in the near future, paving the way for major innovations in areas like smart grids for electric vehicles [7], Industry 4.0 automation [8], critical infrastructure (e.g., gas systems for smart cities [9]), 6G networks [10], the Internet of vehicles [11] or data security [12]. Moreover, such advances can also be enabled by the joint use of other Industry 4.0/Industry 5.0 enabling technologies like Internet of Things (IoT), Edge Computing or Augmented Reality (AR), Mixed Reality (MR), or Virtual Reality (VR) [13].

Previous research outlined potential opportunities for convergence of AI and BC [5, 6]. Nonetheless, as indicated by Pand et al. [14], most current research only provides a theoretical framework to describe the upcoming integration of AI and BC. In addition, most available research focuses on one-way integration (i.e., how BC integrates into AI or how AI integrates into BC) without considering its reciprocal nature, and, in general, it does not take into account the existence of DLTs different from BC that can be more appropriate for specific scenarios like IoT.

This book aims to help AI and BC researchers to develop systems that overcome current challenges and allow the convergence of both technologies.

2. Opportunities and challenges

The full development of BC-AI systems presents numerous opportunities for innovation. Despite the promising foreseen future of such an integration, it is also possible to highlight some open challenges that must be faced by future researchers. The following paragraphs summarize some aspects to be considered:

1. **Current BC and AI maturity levels**

   The adoption of BC-AI systems opens a wide range of potential applications in the short and medium-term. Nevertheless, it is a fast-paced field and there are several potential research topics that would be involved in a full deployment: BC scalability or security, smart contract vulnerabilities, and deterministic execution, trusted oracles election or AI-specific consensus protocols.

2. **Innovation driven by BC-AI systems used jointly with Industry 4.0/ Industry 5.0 Enabling Technologies**

   - IoT. On-device IoT data training is possible thanks to decentralized AI algorithms and mobile edge computing [15]. However, the use of BC in IoT applications still has to face some open challenges [16].

   - Fog and Edge Computing. IoT devices can offload training tasks to fog or edge computing devices to enable AI at the network edge [17]. In addition, such nodes can have a BC module, which allows fog nodes to execute localized data management, access and control. Those distributed edge intelligence frameworks pose advantages in terms of latency and network resource consumption, but impose certain additional challenges such as user privacy and data security [18].

   - Augmented Reality (AR), Mixed Reality (MR), and Virtual Reality (VR) (Metaverse). Such technologies provide new ways to interact with digital content and to create new worlds where it will be possible to engage in different activities that will be accelerated by the use of trusted technologies like BC, thus transforming digital content in valuable assets [19].

3. **Quantum computing**

   - In the coming quantum computing era, new attacks against classic cryptosystems will be developed, therefore researchers will have to pay attention to the quantum computing scene and its advances [20].

4. Reduction of carbon footprint and energy efficiency optimization

- Researchers will have to study how to minimize the environmental impact and maximize energy efficiency [18] when deploying BC-AI systems. Specifically, researchers will have to develop novel approaches to optimize cryptosystems and reduce the energy consumption of AI techniques.

5. Standardization and regulations

- Although standards for BC technology are already currently being developed [21], regulations for BC deployment in the context of AI applications need to be established at local and global levels. In addition, compliance with current legislative directives implies the cooperation of a wide range of global stakeholders and the creation of proof-of-concepts to define the correct set of technical standards for ensuring interoperability.

6. Corporate governance, corporate strategy, and culture

- The ability of organizations to experiment with new business strategies and to make long-term investments will be important in the adoption of BC-AI innovative systems, as a collaborative approach is necessary to incorporate all stakeholders and to establish new ways of creating value while lowering carbon emissions.

3. Conclusions

This chapter summarizes the most relevant issues that will have to be faced by AI and BC during their convergence. Such issues will be addressed in the different chapters of this book, which shows the potential of the integration of AI and BC. Thus, this book includes state-of-the-art and future research opportunities of the convergence of AI and BC.

The book also deals with the technological and practical limitations to be addressed regarding scalability, privacy, smart contract security, trusted oracles, consensus protocols, interaction with Industry 4.0/Industry 5.0 technologies, quantum computing resiliency, reduction of carbon footprint, standardization, interoperability, regulations, and governance. The results obtained from the described analyses will allow for guiding the future developers of interdisciplinary AI and BC applications and/or the convergence of such fields, and to contribute to the development of the next generation of innovations based on BC-AI systems.

Conflict of interest

The authors declare no conflict of interest.

Author details

Paula Fraga-Lamas[1,2] and Tiago M. Fernández-Caramés[1,2*]

1 Faculty of Computer Science, Department of Computer Engineering, Universidade da Coruña, A Coruña, Spain

2 CITIC Research Center, Universidade da Coruña, A Coruña, Spain

*Address all correspondence to: tiago.fernandez@udc.es

IntechOpen

References

[1] PwC's Global Artificial Intelligence Study: Exploiting the AI revolution. PwC, London, UK, 2017. Available from: https://www.pwc.com/gx/en/issues/data-and-analytics/publications/artificial-intelligence-study.html [Accessed: October 24, 2021]

[2] PwC. Blockchain Technologies Could Boost the Global Economy US$1.76 Trillion by 2030 Through Raising Levels of Tracking, Tracing and Trust. PwC, London, UK, 2020. Available from: https://www.pwc.com/gx/en/news-room/press-releases/2020/blockchain-boost-global-economy-track-trace-trust.html [Accessed: October 24, 2021]

[3] Fraga-Lamas P, Fernández-Caramés TM. Fake news, disinformation, and deepfakes: leveraging distributed Ledger Technologies and blockchain to combat digital deception and counterfeit reality. IT Professional. 2020;**22**(2):53-59. DOI: 10.1109/MITP.2020.2977589

[4] Tanwar S, Bhatia Q, Patel P, Kumari A, Singh PK, Hong W. Machine learning adoption in blockchain-based smart applications: The challenges, and a way forward. IEEE Access. 2020;**8**: 474-488. DOI: 10.1109/ACCESS.2019.2961372

[5] Marwala T, Xing B. Blockchain and Artificial Intelligence. 2018. Available from: https://arxiv.org/abs/1802.04451

[6] Dinh TN, Thai MT. AI and blockchain: A disruptive integration. Computer. 2018;**51**(9):48-53

[7] Wang Z, Ogbodo M, Huang H, Qiu C, Hisada M, Abdallah AB. AEBIS: AI-enabled blockchain-based electric vehicle integration system for power management in smart grid platform. IEEE Access. 2020;**8**: 226409-226421. DOI: 10.1109/ACCESS.2020.3044612

[8] Qu Y, Pokhrel SR, Garg S, Gao L, Xiang Y. A blockchained federated learning framework for cognitive computing in industry 4.0 networks. IEEE Transactions on Industrial Informatics. 2021;**17**(4):2964-2973. DOI: 10.1109/TII.2020.3007817

[9] Xiao W et al. Blockchain for secure-GaS: Blockchain-powered secure natural gas IoT system with AI-enabled gas prediction and transaction in smart city. IEEE Internet of Things Journal. 2021; **8**(8):6305-6312. DOI: 10.1109/JIOT.2020.3028773

[10] Li W, Su Z, Li R, Zhang K, Wang Y. Blockchain-based data security for artificial intelligence applications in 6G networks. IEEE Network. 2020;**34**(6): 31-37. DOI: 10.1109/MNET.021. 1900629

[11] Hammoud A, Sami H, Mourad A, Otrok H, Mizouni R, Bentahar J. AI, blockchain, and vehicular edge computing for smart and secure IoV: Challenges and directions. IEEE Internet of Things Magazine. 2020;**3**(2):68-73. DOI: 10.1109/IOTM.0001.1900109

[12] Wang K, Dong J, Wang Y, Yin H. Securing data with blockchain and AI. IEEE Access. 2019;**7**:77981-77989. DOI: 10.1109/ACCESS.2019.2921555

[13] Fernández-Caramés TM, Fraga-Lamas P. A review on the application of blockchain to the next generation of cybersecure industry 4.0 smart factories. IEEE Access. 2019;**7**:45201-45218. DOI: 10.1109/ACCESS.2019.2908780

[14] Pandl KD, Thiebes S, Schmidt-Kraepelin M, Sunyaev A. On the convergence of artificial intelligence and distributed Ledger Technology: A scoping review and future research agenda. IEEE Access. 2020;**8**:57075-57095. DOI: 10. 1109/ACCESS.2020.2981447

[15] Lin X, Li J, Wu J, Liang H, Yang W. Making knowledge tradable in edge-AI enabled IoT: A consortium blockchain-based efficient and incentive approach. IEEE Transactions on Industrial Informatics. 2019;**15**(12):6367-6378. DOI: 10.1109/TII.2019.2917307

[16] Fernández-Caramés TM, Fraga-Lamas P. A review on the use of blockchain for the Internet of Things. IEEE Access. 2018;**6**:32979-33001. DOI: 10.1109/ACCESS.2018.2842685

[17] Fan S, Zhang H, Zeng Y, Cai W. Hybrid blockchain-based resource trading system for federated learning in edge computing. IEEE Internet of Things Journal. 2021;**8**(4):2252-2264. DOI: 10.1109/JIOT.2020.3028101

[18] Fraga-Lamas P, Lopes SI, Fernández-Caramés TM. Green IoT and edge AI as key technological enablers for a sustainable digital transition towards a smart circular economy: An industry 5.0 use case. Sensors. 2021;**21**:5745. DOI: 10.3390/s21175745

[19] Cannavò A, Lamberti F. How blockchain, virtual reality, and augmented reality are converging, and why. IEEE Consumer Electronics Magazine. 2021;**10**(5):6-13. DOI: 10.1109/MCE.2020.3025753

[20] Fernández-Caramés TM, Fraga-Lamas P. Towards post-quantum blockchain: A review on blockchain cryptography resistant to quantum computing attacks. IEEE Access. 2020;**8**: 21091-21116. DOI: 10.1109/ACCESS. 2020.2968985

[21] Anjum A, Sporny M, Sill A. Blockchain standards for compliance and trust. IEEE Cloud Computing. 2017; **4**(4):84-90

Chapter 2

Bitcoin and Ethics in a Technological Society

Eva R. Porras and Bryan Daugherty

Abstract

Bitcoin came into existence as a peer-to-peer payment system for use on online transactions. This achievement was the result of a shared vision about the future relationship between governments' control and citizenry, and the collaborative work of the many who contributed to the development of the cryptographic field. This innovation and its underlying technology, the blockchain, have been at the root of a change of paradigm, as the joint use of blockchain and artificial intelligence (AI) seed the next technological revolution. However, as it is often the case, these revolutionary inventions have also been met with skepticism in the financial sector and society at large. Using the case of Bitcoin and the blockchain, this paper analyzes the intersection between the philosophy and technology underlying these innovations, and the outlook of a sector of society who fears these developments while others try to profit. In this chapter, we first look at the history of Bitcoin together with that of those behind it. We then review the mixed reception it obtained after coming to the market. We assess the innovations' properties and confront these with the needs of a society eager to obtain further clarity and enjoy more transparency in matters of relevance to their participation in democratic processes.

Keywords: Bitcoin, Blockchain, Ethics, Cryptocurrencies, Satoshi Nakamoto, Craig S. Wright, Distributed Ledger Technologies, Artificial Intelligence, Data Privacy, Freedom, Crypto-anarchy, Libertarianism, Cryptography

1. Introduction

" On ne résiste pas à l'invasion des idées." – Victor Hugo, Histoire d'un Crime

Bitcoin [1] came into existence in 2009 as a peer-to-peer payment system for use on online transactions. This type of electronic cash system was designed to make online payments without the need of an intermediary financial institution to coordinate the transaction. The system became known when an individual using the pseudonym of Satoshi Nakamoto broadcasted the first version of the protocol in October 31st 2008 and released the related software in January 2009 [2, 3]. This software could be downloaded by anyone, and any computer running it could join the network. With Bitcoin, third parties to a transaction become dispensable because now the exchanges could be executed with no middleman to connect the sender and the receiver. Instead, the operations used a network of computers that communicated with one another directly through the Bitcoin open source software.

By the time Bitcoin was introduced to the market, there had been multiple prior attempts to launch a digital currency (E-gold, or the Liberty Reserve). Given the long history of technological evolution prior to the public coming out of Bitcoin, we can say that the ideas that led to its success were in the making for decades. This achievement was the result of a shared vision about the need to develop electronic payment systems in a way that was coherent with the concurrent evolution of online technologies. In addition, this outcome responded to the wants of an ultra modern society in which the needs for privacy, efficiency, effectiveness, and transparency could not be ignored and would have to inhabit shared spaces with a perception of individual freedom. Thus, although Bitcoin was masterminded by Satoshi Nakamoto, this technology was the result of the active collaboration among individuals who, each in its own way and to different extends, cooperated in its development.

However, aside from many other conceptual and technical differences, in contrast to the earlier failed attempts at creating various types of digital cash, Bitcoin became successful and it remained the only decentralized digital cash coin until Namecoin emerged in April 2011 [4] as the first "alt coin". A key reason for this success, is that the creator of Bitcoin was able to incorporate in the design the solution to two long-standing conundrums: the *Double-Spending Problem* and the *Byzantine Generals Problem*. The first of these problems refers to ensuring the information received is complete and accurate and no falsified updates get introduced into the ledger so that the same money is never spent more than once. The second problem relates to the reaching of consensus among parties who do not trust each other because they do not share the same interests.

Ever since becoming public in 2009, the "block chain" or "time chain" -as Nakamoto first called Bitcoin's underlying technology -, has been at the root of a change of paradigm. The reason is that the joint use of the blockchain and AI is expected to seed the next technological revolution. This is so much the case, that a new economic sector has already surged around engineers and inventors who are developing applications in various industries.

Together with this technological expansion, the hype of a revolutionary development and, particularly, the promise of huge potential economic rewards has also brought herds of people into performing other activities around these new sectors. Some have become entrepreneurial miners, and others have gone into performing roles such as those of investor, trader, and/or speculator of these markets. These events have occurred while the world at large has taken the role of spectator: on the one hand attempting to capture the essence of this technology, and on the other hoping to envision what, if any, could be the potential uses and the consequences for society of its meaningful implementation. Concurrently, the speculative nature of the financial markets around most of these assets has become undeniable and worrisome.

One essential impediment preventing the fair evaluation of the various solutions grouped now under the general umbrella of "cryptocurrencies" or "alt-currencies" is the technical complexity of these products. This explains why many publications and investors mistakenly compare and think of them as equivalent. In addition, there is also an underlying intellectual and moral battle among those who do understand the technology as to what attributes should define their structure and substance. In particular, there is the key issue of traceability, one that was already at the core of the evolutionary history of the creation of electronic payment systems.

Beyond that, as it is often the case with innovations, these have raised strong emotions among many in society. Part of these emotions are explained by the challenges presented when trying to adapt to the existence of the new technology. But, in addition, much of the emotional tide surges due to the issues highlighted in the prior paragraph. The lack of understanding of how the technology works and the permeability of attitudes rooted on moral grounds have resulted on high-peach-statements by many including relevant figures in the financial sector and society at large. The following are three early examples of the animosity of relevant public figures who use skepticism and express abhorrence at the new technological revolution. These are: "Bitcoin Is Evil [5, 6] "by Nobel award winner Paul Krugman; "Why I want Bitcoin to die in a fire [7] "by Charlie Stross; and multiple declarations by JPMorgan CEO, Jamie Dimon [8, 9]. Given that historically Bitcoin is the most recognizable and relevant among the assets grouped as "cryptocurrencies" much of the criticism uses it as a representative of the asset class.

2. The deep web and cypherpunks

Independently of the differences among the various digital assets, cryptocurrencies -as an asset class, and the blockchain -as a technology, have awaken strong emotions in market observers and participants. At the heart of the problem is whether these technologies merit their own existence; and if so, how to house them within the common categories of property and personal rights. Beyond that, the early use of these technologies by individuals in the deep web to make illicit and illegal trades [10], casted a negative shade that has proven difficult to shake. This negative impression has been further cemented by a general unscholarliness about the workings of these technologies and the inability of the common reader to tell these apart from each other. For instance, there is a generalized understanding that Bitcoin is "untraceable digital cash." As such, this digital cash tool could be potentially used to avoid the payment of taxes and to finance a myriad of illegal activities such as drug trade, terrorism, kidnapping, and extortion. So the semi-anonymity or anonymity quality of many crypto currencies is at the core of this unfavorable perception. Nonetheless, this sentiment is entrenched also due to the legend of the cypherpunk movement.

The cypherpunk was a 1970s' movement that advocated for less government control which, in their view, stifled economic growth [11]. This belief came together with a libertarian notion of freedom, and the intuition that a strong cryptography could guard against government interference in personal matters [12]. One aspect of the objectives embraced by this group, dealt with restructuring how people economically interacted with one another. And the solution proposed was the use of a digital cash currency that would be free from government control. In his paper "b-money [13]" Wei Dai described:

> [] A community is defined by the cooperation of its participants, and efficient cooperation requires a medium of exchange (money) and a way to enforce contracts. Traditionally these services have been provided by the government or government sponsored institutions and only to legal entities. In this article I describe a protocol by which these services can be provided to and by untraceable entities.

Centered around the Cypherpunk email list [14], the group championed encryption as a way to shift power from the government to individuals. And as public-key cryptography evolved, they began to conceive how a future society could deal with money. Their attempts to develop a digital cash currency that would be free from government control underwent numerous stages and, through time, various publications described the possible structure of this future cash. However, it was David Chaum, the one who first proposed digital cash as files of digital value that were anonymous and exchangeable [15]. His 1981 paper: *Untraceable Electronic Mail, Return Addresses, and Digital Pseudonyms* [16], was the cornerstone for later research of "anonymous communications".

The cypherpunk generation achieved great progress towards the development of a decentralized, strong, online currency. For instance, Chaum created an algorithm which allowed the modification of coins without breaking the signature of the mint. In his 1982 paper "Blind signature for untraceable payments. [17]" Chaum explained that the growth of electronic banking services, and the creation of automated payment systems would require to balance the need for personal privacy and the potential for the criminal use of payments. He then summarized that the ideal payment system would have the following three key properties:

1. Inability of third parties to determine payee, time, or amount of payments made by an individual;

2. Ability of individuals to provide proof of payment, or to determine the identity of the payee under exceptional circumstances; and

3. Ability to stop funds which have been reported stolen.

To illustrate the use of this technology, Chaum proposed how by fulfilling the three enumerated properties electors at an election event could vote without having to meet at the electoral school to drop their secret ballots. Chaum's system would balance the need to keep the vote secret, the ability to verify that the vote was counted, and the capacity to prevent voter fraud.

In addition to Chaum, several pioneers also worked in other versions of electronic cash. One example is Hal Finney, a developer that came out with a Reusable Proof-Of-Work (RPOW), a short-lived solution called this way because it was based on proof-of-work [18]. But, as said, it was not until 2009 that after decades of technological evolution, hard work in cryptographic research, and many failed attempts, Bitcoin came into the market to become the first digital cash coin capable of withstanding the process of its own development.

3. Emotions versus facts, and perception versus reality

There are many reasons why people resist change. For one, change is a psychological experience that requires time to process. Furthermore, if change is big and unexpected a common reaction is denial. In this scenario, we can tell ourselves that nothing of relevance is happening and excuse our participation in the process. Feeling unprepared for the new environment also explains this rejection as people are pushed out of their comfort zones. Change implies a departure from the "old ways." Hence, those who did not catch up to the new version might feel superseded and are bound to be defensive about it. And, if change involves a new technology, a common concern is personal competence. People worry that their skills will be obsolete and, as a defensive mechanism, they might express skepticism about the success or adequacy of the innovation. In addition, change is likely to imply more work and this may ripple into resentments and other negative feelings. At the end, depending upon the position of those affected by it, resistance to change may be externalized in one of a variety of manners, from foot-dragging behavior and indolence, to sabotage and rebellion.

When new technologies displace old ones it appears as if whole sectors of society will be hurt. This will be particularly true when those affected resist catching up with the times. In these instances, the damaged sectors can be quite large, as they might include different industries such as providers and users of the old technology. The emotional experience of these processes of resistance has been compared with "being irrational" (see Fineman, 1993 [19]). From this point of view, emotions are understood as the root of the problems, rather than an expression of the underlying difficulties confronted during the implementation of change. From a psychological perspective emotions are not necessarily destructive as they help individuals adapt to difficult situations. But they might motivate an unhealthy resistance that can block the ability of those under stress to assess the situations properly.

The common reactions to change introduced in the prior paragraphs might be able to explain, in part, the strong emotions shown by mainstream media outlets and many relevant figures in society when reporting about cryptocurrencies. Albeit the recentness of the innovation, and that the high volatility experienced in these markets alerted many, the way concerns were expressed frequently showed a high level display of emotions as well as a limited understanding of the technology. These expressions of "hate" most often addressed all products grouped under the "cryptocurrency/Alt-currency" headings as if these were equal or equivalent assets. That is, in general, many commentators did not differentiate between key aspects of the technologies underlying these assets. Be it news, investment or entertainment, television or written press, online media including social media outlets, all expressions published on these forged the vantage point of millions of people when thinking about cryptocurrencies.

For instance, at a public forum reported by the Financial Times on February 15, 2018th, Berkshire Hathaway vice chairman Charlie Munger depicted Bitcoin as "totally asinine [20]" adding it should receive a government crackdown. On March 5th, 2018th, Harvard economist Kenneth Rogoff told CNBC reporters that Bitcoin is *"more likely to be worth $100 than $100,000"* by 2028 implying its value depended upon its use in *"money laundering and tax evasion* [21]." Another laud Bitcoin basher was JPMorgan CEO, Jamie Dimon [22, 23] who declared publicly and repeatedly his disdain for Bitcoin. For instance, during a public conference in New York, Dimon declared that trading the virtual currency *"was stupid"* and he [24] would *"fire in a second"* anyone found doing it at his firm. Later, while at the Aspen Institute's 25th Annual Summer Celebration Gala on August 5th 2018, Dimon called Bitcoin a *"scam"* and a *"fraud* [25], "and reiterated comments he had made a year earlier when stating that Bitcoin was:

> *"worse than Tulip Mania"* and *"only for people in countries like Venezuela [26, 27], Ecuador or North Korea [28, 29] "or a bunch of parts like that, or if you were a drug dealer, a murderer, stuff like that, you are better off doing it in Bitcoin than US dollars." "So there may be a market for that, but it'd be a limited market."* He further argued that *"governments should shut them down if they were uncapable of controlling them"* [30–32].

Three examples of articles whose titles already show laud emotional content are: 1) Nobel award winner Paul Krugman's "Bitcoin Is Evil [33]"; 2) Charlie Stross' "Why I want Bitcoin to die in a fire, [34]" and 3) Nobel Laureate Robert Shiller's "Cryptocurrencies have a mysterious allure – but are they just a fad?" [35] Some of the statements made in these literally include: *Bitcoin comes with an implicit political agenda attached, it is designed to be untraceable, and easy to hide, libertarians love it because it pushes the same buttons as their gold fetish and it does not look like a "Fiat currency", it will become central to a commodities markets where the goods traded will include assassination, drugs, child pornography and so on, Bitcoin was designed for tax evasion, Cryptocurrencies are designed by people who hold themselves above national governments* [36].

Given the histrionic nature of many publications, in 2018 Gareth Jenkinson developed the idea of testing the waters of "hate-going" emotions when it came to cryptocurrencies. His findings were published in a cointelegraph article: *Tulips, Bubbles, Obituaries: Peering Through the FUD About Crypto* [37]. In this work, the author showed that during the nine-year-existence of Bitcoin, more than a handful obituaries asking for its 'death' had been published. These writings came from a wide variety of industry experts and commentators who offered their overall subjective and negative comments, showing a fear-mongering mentality that tried to belittle the breakthroughs sparked by the blockchain technology. In his section "A brief history of Bitcoin deaths," the author analyzed instances when mainstream media outlets had signaled the death of Bitcoin. By 2017 these obituaries [38] contained 118 articles. As of mid-January 2021, this figure had increased to 395. Their conclusions were based on assumptions or quotes from a wide range of commentators who used fraud, money laundering, Ponzi schemes, and the likes to announce Bitcoin's demise. A glance down the list of headlines from the various publications helps assess the profound effect these could have had on the sentiment of many people. The examples brought here referred to Bitcoin but this type of press also affected other crypto assets [39] such as Ethereum. In this case, it was the web Digiconomist [40] the one who compiled the list of Ethereum obituaries between 2015 and 2017. Criticism has also affected other cryptocurrencies with plenty of pessimistic forecasts.

With respect to some of the most common criticisms, rebuttals have used the following arguments [41]:

1. the ironic weakness of fiat currencies [ie. the dollar has lost 98% of its value over the last 100 years [42] or

2. the fact that it was it JP Morgan, rather than Bitcoin, the one that was bailed out by the government [43] in 2008 at the tune of US$25,000,000,000 from the Troubled Asset Relief Program (TARP) program [44] while admitting they did not need the funds, or

3. the fact that his bank has only 10% of what they claim to keep from deposits, and uses "Fractional Reserve Banking" to create 90% of its money out of thin air every day, but still claims it is the bitcoin currency the one that is illegitimate and the one with the real problem the government must stop, or

4. the fact that he is a CEO of a bank who for years sat on the Board of Directors for the Federal Reserve Bank of New York, the one that regulates his bank, or

5. the fact that J. P Morgan Sees Crypto as 'competition' and 'risk' as it was stated in the *"Risk Factor"*, segment on cryptocurrencies, of their 2017 Annual Report to the Securities and Exchange Commission (SEC) filed Feb. 27 [45–47].

With respect to the latter one, this report uses the generic "cryptocurrencies" under the *"Competition"* subsection of Item 1A of Risk Factors to explain a change in landscape with new competitors that can threaten J.P. Morgan's operations:

> *"Both financial institutions and their non-banking competitors face the risk that payment processing and other services could be disrupted by technologies, such as cryptocurrencies, that require no intermediation."*

The new technologies

> *"could require JPMorgan Chase to spend more to modify or adapt its products to attract and retain clients and customers or to match products and services offered by its competitors, including technology companies." And eventually this competition could "put downward pressure on prices and fees for JPMorgan Chase's products and services or may cause JPMorgan Chase to lose market share."*

These observations are not farfetched as competitors have come to realize the potential of cryptos. This became particularly obvious as fellow giant Goldman Sachs revealed it was looking into the creation of Bitcoin Futures [48, 49], planned to buy and sell cryptocurrency and offered various contracts with Bitcoin exposure [50–52]. According to Goldman executive Rana Yared: the bank is not a bitcoin believer but it had to acknowledge multiple clients' requests to work with bitcoin.

Goldman and JP Morgan are just two among many banks who are taking notice of the changing environment. For instance, in its annual report to the Securities and Exchange Commission (SEC) filed 2018 Feb. 22 [53] Bank of America (BoA) stated to feel behind and "unable" to compete in the growing crypto market. In this report BoA recognizes that it will have to afford major costs to remain competitive in the cryptocurrency arena [54, 55]:

> *"Our inability to adapt our products and services to evolving industry standards and consumer preferences could harm our business," BoA states in the filing: "the widespread adoption of new technologies, including internet services, cryptocurrencies and payment systems, could require substantial expenditures to modify or adapt our existing products and services []."*

Thus, BoA decided to innovate by requesting a patent for a cryptocurrency exchange system. However, this has not prevented the bank from stopping their clientele credit card purchases of cryptos [56] as the bank is very aware of how the new competition will be detrimental to its prospects as read in their SEC report [57]:

> *"...The competitive landscape may be impacted by the growth of non-depository institutions that offer products that were traditionally banking products as well as new innovative products," and "this can reduce our net interest margin and revenues from our fee-based products and services. In addition, the widespread adoption of new technologies, including internet services, cryptocurrencies and payment systems, could require substantial expenditures to modify or adapt our existing products and services []."*

BoA's declarations to the SEC as well as those made by other institutions such as JP Morgan Chase [58] recognized that while cryptocurrencies endanger their business they are *"innovative"* and *"unlikely to disappear"* as they note obvious advantages in several traditionally problematic or slow areas, such as cross border or international payments.

With respect to Krugman's statements [59], his opinions have been challenged [60] on the grounds that the Bitcoin technology is an electronic payment system designed to work directly between sender and receiver, thus saving the users the 2–3% or higher tax taken by the processors. As a payment system, this technology is ethically neutral even if some use it for unethical purposes. Equivalently, the banking system has also been used by many to make illegal payments, but we do not call HSBC or the Deutsche bank "evil". Hence, one could interpret that Krugman's opinions, these and others, are built on an emotional defense of the status quo: the central banks, payment intermediaries such as Visa or MasterCard, and the State in general.

Virtual currencies are just a form of private money that uses Blockchain technology to record the transactions. But this technology can be easily built upon to address problems and gain efficiencies and effectiveness in multiple types of operations, so its potential uses across industries are boundless. Also, the transaction networks are comparatively safe, transparent, fast, and borderless. So economists who try to belittle and discredit their relevance on the grounds that these are concoctions of "quacks and cranks" (Skidelsky, 2018 [61]), tools for money laundering, crime, and tax evasion, or a renewed version of old libertarian or bubble manias (Shiller, 2018), are simply wrong. For instance, it is easy to clear up two of the most common misconceptions:

CLAIM: Bitcoin's main use is laundering money and making payments for illegal trades.

REBUTTAL: Bitcoin's underlying technology is the blockchain: a ledger that keeps a permanent record of all transactions ever made since the beginning of its existence. This permanent record registers every holder (a bitcoin wallet) of each coin. So the records tie each bitcoin with one or more wallets. The wallets are handled from smartphones and computers so even though technically the bitcoins are not associated directly, nonetheless, they are associated indirectly with a person through the electronic device. Thus, illegal transactions can be spotted, cannot be erased, and can be tracked to a specific individual.

CLAIM: Bitcoin helps avoiding taxes.

REBUTTAL: As already said, every bitcoin transaction is permanently recorded and publicly accessible. Thus, the Internal Revenue Service (IRS) or any equivalent organization can track their movements and easily estimate any taxes due for any individual.

4. Ethics in social networks

Unfortunately, current governance within the social networks does not help distinguish legitimate sources from others trying to piggyback on their work. And apparently, there is also a problem actively prosecuting people and companies who use the image and name of others to confuse and lie to the unsuspecting visitor. This is true at various levels: the private corporation having a direct responsibility over their networks and actions, as well as the government level having the authority and responsibility to ensure corporations can grow within the rule of law.

During a recently recorded conversation with Ryan X. Charles, Dr. Craig S. Wright provided one such case as an example [62]. The situation he relates, refers to a complaint he had placed on Twitter. The motive was to bring their attention to copyright breaches under the Digital Millennium Copyright Act [63], and asked the network to take action against people using copyrighted photos of himself. In response to his request, Twitter deplatformed Dr. Wright while, apparently, taking no action against those using the copyrighted images, which could be found posted at the network thereafter. But, Twitter is not alone. For instance, just as recently as January 26th. 2021, numerous accounts using Dr. Wright's picture and name could be found on Instagram.

In the Abstract of his January 2021 work "An exploration of ingroup behaviour and social psychology in developing socially abhorrent behaviours in social media and financial systems" Dr. Write summarizes the following related observations [64]:

> *"The following paper provides a preliminary investigation into the growth of "Cryptocurrency" subgroups, the abuse of social media using automated systems, the enhancement of trolling and the ability for these activities to pose both a political and financial threat. Malicious actors have utilised technology to leverage existing psychological behaviours and create tribalistic responses that allow for the automated approach to controlling and manipulating individuals online. In this, authoritarian leaders can asymmetrically leverage sociological and psychological benefits that developed through evolutionary benefits but yet exhibit adverse effects in modern societies."*

and concludes [65]:

> *"The ability for malicious actors to use anonymous social systems and technology has allowed for the creation of criminal groups that target political systems, financial systems and generally cause dilemmas that result in lost economic opportunities for*

many people and may even go as far as Social psychology causing personal harm. In providing access to a wide variety of platforms that can be tied to fake and manipulable sources such as those controlled in asymmetrical systems using bots, authoritarian and socially deviant actors can manipulate others to polarise and partisanise groups. These results may be seen in the false manipulation of Cryptocurrencies including Bitcoin through groups such as BTC Core and the introduction of specialist language for ingroups who believe not only that they will get rich, but they will gain in power and prestige. Consequently, the rise in new technologies that allow for the disassociation of the individuals' identity and the creation of methods that allow individuals to distance themselves from their activities must be investigated to regulate these systems".

5. How can we judge what is ethically right?

In his Tanner Lecture *Science and Revolutions* [66] Sagdeev states that "the intellectual community rarely has been the direct beneficiary of revolutions". In his words, this group "has played the role as a patient, the victim of change; and as a doctor, preparing and implementing the revolutionary processes [67]". The truth is that, even though Sagdeev is referring to other types of uprisings, Bitcoin and the Blockchain are providing a comparable revolutionary environment. This revolution too comes with a conflict of interest: on the one hand the political slogans, the power plays, the status quo, and those taking advantage of the confusing environment to loot for their own personal benefit, and on the other the intellectual drive to search for truth, rationality, and progress.

The attempt to use intellectual thinking to social political phenomena is, in Sagdeev's view, one reason why scientists become the "first revolutionaries, and often the first prisoners after the success or failure of revolutions [68]". As an example, he cites the time Einstein and a group of physicists became victims to the attacks of Soviet philosophers who demanded quantum mechanics be liberated from the: ""bourgeois" principle of uncertainty," and the theory of relativity be "liberated" from the dubious role played by imaginary observers [69]".

In those times, science was hostage to the "supreme wisdom" of communism as given in the form of proclamations by the classic manifestos of Marxism. And many of those incapable of undergoing the soul engineering process required to produce the "new Soviet man" or hiding successfully, were exterminated. Comparable events had been experienced at earlier revolutions, such as the one in France when Jean-Paul Marat demanded that chemistry in particular, be a "people-friendly science." This resulted in a general bloodshed including the beheading on the guillotine of the founder of chemistry: Antoine Lavoisier, whose ideas about the nature of chemistry differed from those Marat had.

Now, we live in different times. But still we can feel a serfdom, not subservient to a recognizable regime, but rather to a plethora of forces, —be some big corporations, be a number of governments, be the concept of the welfare state, be other sources of status quo power such as communication giants- that also try to mold at their convenience a type of "new modern man" and determine who is the worthy intellectual. Against these attacks, each person can chose to go the way of "internal emigration" and keep quiet, just as in the old Soviet Union, or face concerted efforts to end with one's prestige and reputation and maybe even ones' physical safety [70].

As things stand now-a-days, it does appear that our current "intelligentsia" will also need to split into at least two groups. In Spain for instance, the one who wants to progress, might be forced to be de facto "above their own national government" such as Shiller suggested [71], in search of a milieu where growth is not stopped. The reason is that in the agenda of the civil servants and politicians responsible for ensuring legislation catches up with the reality of the times, this is not a priority. However, Bitcoin and blockchain-based innovations need a regulatory system that is flexible, clear, transparent, agile, and competent. And of course, that would require those regulating the environment understand the technology. One more reason the responsible agencies and business groups should clear the air as to what is true and what is not.

Obviously, when entrepreneurs cannot obtain the necessary licenses or the processes are delayed in such a way that their inventions become obsolete, or they suspect their fiscal obligations might not be clear, they are forced to rethink their situation. In Spain, he following is a list of some of the problems faced by managers wishing to organize these businesses according to the law [72]:

1. Obtaining electronic money licenses takes a long time.

2. Given the authorities' limited understanding of the technology which they view as a financial asset rather than a protocol, there is also uncertainty on how to treat tokens for the purposes of taxes.

3. The same goes with respect to the possibility and agility to enter the legal sandbox. In Spain the sandbox has just been approved, and it remains a bureaucratic and administrative mess.

4. Companies cannot set up and manage their firms 100% remotely, without some face-to-face activities. This requirement boosts set-up costs and adds no value.

5. Company taxation and administration is more expensive than other jurisdictions such as the US, the UK and Switzerland, which in addition offer fewer obstacles to growth.

These are just some of the key problems confronted by digital-cash and blockchain entrepreneurs who want to set up their businesses within the Spanish territories. Much of the void can result from the government's belief in Bitcoin as a financial asset or financial instrument, rather than a communication protocol and therefore they continue to believe that Bitcoin is for speculation and for criminals When an environment is not legally and technically ready to receive innovations, the credibility of those set under that administration suffer. In the case of our example, the many relevant and significant inadequacies of the Spanish system force many local innovators to leave their country and set up their companies abroad, mainly in Switzerland, the UK, and the US. That is because these countries have managed to develop a more transparent and user friendly environment easing both the rate of company creation and the rate of technological transfer into their borders.

This situation raises legal and ethical problems for all involved but in addition, it also has strong financial consequences. However, often, in lieu of fixing the problem, governments such as the one in Spain try to stop the rate of development by creating a bureaucratic maze or by confining its development within organizations they control directly or indirectly. Once more, this situation brings to mind the environment in the Soviet Russia where most scientific development was scheduled by the political authority and supported by work on contracts or grants.

6. The honest truthful asset

In the first part of his 1986 Lecture titled *How Is Legitimacy Possible On The Basis Of Legality?* [73], Jürgen Habermas questioned whether "legitimacy" is possible on the basis of "legality". And to highlight the conflict and incongruity hidden within this statement, he used Max Weber's vision of Western political systems as forms of "legal domination [74]." The point being that the legitimacy of these political systems resides on the belief in the "legality" of their exercise of political power versus, say, that of the "tradition".

In current modern democratic societies, the acceptance of such a premise may create contentions that cannot be resolved within the existing political structures. The reason is the conflicts of interests inherent to such systems. For instance, the most important objective of a political party in a democratic system is to be reelected. And, to achieve this end, politicians will often use public assets, such as public mass media communications, as if these were their own. Given that all political parties share the same interests and thus will benefit from these actions, checks and balances may be removed so each of them can take turns at abusing the system. Furthermore, given that the underlying infrastructure of the "welfare state" consists in taking wealth from some sectors of society, using a part of these to support the apparatus, and redistributing the remainder among other groups expected to became the captive electorate of the parties in power, we can already see situations when the rights and property of first are threaten to benefit the latter.

On October 2015, acting as the moderator of an "All-Star Panel" during a Bitcoin Investor Conference at Las Vegas, Michele Seven asked about the nature of property rights [75]. The first to answer, Dr. Craig Wright [76], highlighted:

" [] We need to be able to control our own freedoms and the only way to do that is to basically have the right to property, to ownership [] That means being able to dispose of property as we want, to be able to share it, to take it -- and that is what it is all about. Once we get things to where we have redeemable contracts and we link them to the blockchain, where we can link money, and goods, digital rights, and ownership into something that can't be changed: a fundamental open, honest, truthful asset -- the blockchain, that's when we are going to see real freedom in the world."

With respect to the same question, Joseph Vaughn Perling [77] reminded the audience that currently one relies on government ledgers to keep property records which tell us who owns what, and that such system is unreliable and expensive as it is financed through taxes. Nonetheless, with the new technology, all these costs can potentially be reduced as property records get stored in the ledgers of Bitcoin. Then, reflecting on the potential future conflict of interests between society and power centers, Joseph Vaughn Perling [78] added that there may come a time when:

> *"the separation between the honest politician and the dishonest one will come down to whether or not they support the use of Bitcoin for government function because it does provide that audibility and the anti-corruption tools that it can implement throughout. Government can make government become provably honest in a way that's never before been possible and provably honest government is something we have never seen" [] so that it may create that division between the people: the people within government that become more electable because they can prove the degree to which they are honest, versus the those who are competing for their office."*

The use of blockchain to secure a more transparent political arena will be an interesting development particularly in light of the practice of "legal domination" by which the rationality that the law possesses, is independent of morality. Now-a-days it is impossible to imagine a society where citizens do not demand that it is the moral argumentation that gets institutionalized by means of legal procedures. And this expectation will need to materialize results over all aspects of government including those that impact science.

Baumol (2002 [79]) stated that *"virtually all of the economic growth that has occurred since the eighteenth century is ultimately attributable to innovation."* Given that the blockchain is thought of as, probably, the most auspicious innovation since the coming of the internet, this invention is prophesied deliver huge financial benefits. These will result from the economic repercussions of its incorporation into processes to streamline and secure decentralized transactions in countless sectors across the world. The blockchain is specially relevant to situations when ownership histories are of essence such as in the pharmaceutical industry, land registries, real estate property, piracy and copyright matters, as well as of public services, such as health assistance and welfare payments (Tapscott et al., 2016 [80, 81]). In the limit of this innovation are self-executing contracts that can run with the assistance of AI and minimal human intervention. The use of the blockchain will provide increased efficiencies and more cost-effective solutions to current predicaments. And as the older technology is replaced, the blockchain will reduce fraud increasing trust and security, and it will improve the transparency of multi-party transactions.

Given all of these, one would expect public institutions would align to welcome and assist to facilitate the said developments. However, in the current atmosphere of political and economic deterioration, where political and status quo agendas control the rate of development, the scientific community and the entrepreneurs who are willing to finance these are at a loose in a rather hostile psychological climate.

7. Restoring trust, transparency and efficiency in government with a publicly scaled blockchain

In a letter to W. T. Barry, on August 4th 1822, James Madison [82] stated:

> *"A popular Government, without popular information, or the means of acquiring it, is but a prologue to a Farce or a Tragedy; or perhaps both. Knowledge will forever govern ignorance: and a people who mean to be their own Governors, must arm themselves with the power which knowledge gives."*

Transparency and accountability are of the two most essential principles in a free and democratic society. They are a bridge between an informed citizenry making confident electoral decisions or the widespread distrust of 'a self-serving, arbitrary, corrupt institution'. Furthermore, transparency and accountability are ever so more important as corruption keeps eroding the legitimacy and credibility of democratic governments worldwide. According to Pew Research Center, in the US public distrust of the government and elected officials has eroded to reach all-time lows [83]. This has been highlighted with the rise of civil unrest, violent protests, and frequent demonstrations against government policies, politicians, and media organizations [84, 85]. The erosion of public trust in government and news media can be attributed to numerous factors, many of which relate to the honesty, openness, and confidence, or lack thereof, in the information that is disseminated.

In 2011, the US launched a comprehensive digital government strategy aimed at building a 21st century digital government [86]. The Executive Order highlights:

> *"Government managers must learn from what is working in the private sector and apply these best practices to deliver services better, faster, and at lower cost. Such best practices include increasingly popular lower-cost, self-service options accessed by the Internet or mobile phone and improved processes that deliver services faster and more responsively, reducing the overall need for customer inquiries and complaints. The Federal Government has a responsibility to streamline and make more efficient its service delivery to better serve the public."*

However, ten years later we can still find proof of the Government's slow response to technological shifts. For instance, on December 2020, the Cyber-Security and Infrastructure Agency revealed that a yearlong hack had affected US private firms, government agencies, and critical infrastructure entities [87]. These included: the US Treasury, Department of Homeland Security, Department of State, Department of Defense, Department of Commerce, National Institute of Health, Center for Disease Control and Prevention, and the Justice Department among countless others. In total, it is estimated that 18,000 entities fell victim to the Russian hack. This relatively unknown hack is expected to cost American businesses and taxpayers over $100 billion dollars [88]. These types of attacks targeting the common citizen are so frequent that the ethical aspects of these actions blur against all the other consequences of these scandalous activities which ten-fold with time.

7.1 Building trust with bitcoin: now is the time for a blockchain reformation

Similar to the transformative nature of the internet, a public blockchain has the ability to revolutionize government processes by providing greater transparency and auditability as well as a super-efficient "Universal Source of Truth" data management platform that can be used to restore trust, authenticate data, and significantly reduce costs. This has become true after the publication of the 2008 Bitcoin Whitepaper by Satoshi Nakamoto which presented solutions to long-standing issues such as the scaling obstacles among other [89].

Bitcoin was designed to be the foundation for an open and honest system, one that is public, has a series of checks and balances, as well as an incentive for participation based on Proof-of-Work. On the Bitcoin network, every transaction is recorded on a public ledger maintained by a small-world network of specialized distributed nodes called transaction processors. As transactions are broadcasted, processors gather, validate, timestamp, and add each transaction as it is received in a series of hash-based, agreed upon chain of events, secured in blocks of immutable information.

As explained earlier, contrary to much of the popular belief, Bitcoin offers more than just a transfer or store of financial value. Bitcoin establishes a Universal Source of Truth, where information can be stored, validated, shared, protected, and authenticated. This can be used in conjunction with traditional systems or new hybrid options utilizing cloud to chain solutions. Not all data has to be stored on chain, but rather information can be authenticated simply by hashing it in the cloud and storing a copy of that hash on-chain. This would ensure that the data stored in the cloud, or elsewhere, could be simply authenticated to confirm it has not been changed.

Restoring the Bitcoin original protocol by removing the real centralization bottleneck, has allowed true innovation and unbounded on-chain scaling to occur. On May of 2020, the Bitcoin SV blockchain processed a world-record size block of 369 Mb which contained 1.3 million transactions. In fact, the network has already eclipsed almost 4,000 transactions per second (tps) and is expected to reach 50,000 tps later this year. Through scaling comes cost efficiency and Satoshi's vision remains unmatched in its ability to transfer micro – even nano-transactions with a median transaction fee of 1/100th of a U.S. cent. With safe, instant, low fee transactions of Bitcoin SV, government organizations can significantly reduce costs associated with financial and data transactions. These savings may be compounded by a reduction in the associated costs with auditing, cybersecurity, and networking hardware.

7.2 Bitcoin can help governments restore transparency and trust today

Bitcoin SV stands ready to fulfill the promises of an era of blockchain refor-mation by providing complete transparency and efficiency to the public sector. Although government entities would only need to begin with a common Request for Information, traditionally, the procurement stage has long been considered one of the greatest barriers to connecting government technology needs with vendors who are able to integrate the latest emerging technologies [90]. The consequence is that many small firms and industry outsiders are shut out entirely from participating due to how complex, time consuming and costly the process can be. In contrast, a myriad of transformative blockchain solutions await to contribute to a more ethical society by improving transparency and restoring trust. Some of these are:

Financial Transaction Management – As a distributed ledger, Bitcoin offers an accounting of valid transactions that occur within the network instantaneously. For a small transaction fee (.00011 per byte), transaction processors will record an entry onto the secure ledger. Compared to the cost of modern transaction management systems, Bitcoin offers unmatched savings, auditability, security, and interoperabil-ity with the integration of smart contracts and tokens. *Example:* **Tokenized** [91]

Regulatory Compliance – As transactions are validated and publicly recorded to the Bitcoin blockchain, they are secured by an immutable Proof-of-Work. This allows regulators, news media and government watch-groups real-time access to compliance-related data that can be shared and trusted. In return, this eases the burden of reporting and auditing on government agencies, reducing cost and improving transparency. Smart contracts for government procurement opportuni-ties would ensure compliance, fairness and improve the overall speed of implemen-tation. *Example*: **nChain** [92]

Identity Management - Unlike centralized government databases, Bitcoin provides a much more secure distributed data management platform that could empower citizens with the ability to easily sign and authenticate their identity for official government documents or benefits. This would also reduce the time and resources needed by the government to verify identities and protect sensitive data – especially across restrictive inter-agency data silos. *Example:* **Legally Chained** [93]

Registries – The ability to manage any type of record or registry through Bitcoin's unique data management network removes the overall complexity for governments to manage and authenticate data efficiently. This would remove the friction of processing land titles, company registrations as well as every other type of record including birth, marriage, divorce, criminal or death. The ledger would serve an honest, universal source of truth that can drastically reduce fraud and corruption. Example: **Elas Digital** [94]

Blockchain Voting – As we saw during the 2020 US Presidential election cycle, it is important for citizens to believe in the integrity of the voting process. Doubt in returned results, whether due to error, fraud, hacking, corruption, or lack of trans-parency can create an atmosphere of distrust among voters. Bitcoin's tamper-proof public ledger is perfectly suited to eliminate election fraud in the future – when combined with an identity-based token, a voter could easily cast their vote using any type of device removing barriers and increasing participation. Example: **Layer2 Technologies (B-vote)** [95]

Supply Chain Traceability – The Coronavirus pandemic has demonstrated to all how fragile our global supply chain can be during a disruption. Government agencies competed to locate, purchase, and distribute medical gear, supplies and personal protective equipment. This created panic among the populace as medical care was either denied, delayed, or compromised through the reuse of protective gear. The lack of traceability of the supply chain continues to plague COVID-19 relief. As traditional government vendors begin developing vaccine distribution and contract tracing technologies, many citizens are concerned about how their personal medical data will be stored and used in the future. Bitcoin solves these issues by improving trust and privacy among parties that need to share valuable data across an entire value chain. Example: **UNISOT** [96], **VXPass** [97].

- **Health Care** – A public health crisis like the Opioid epidemic carry a heavy cost on communities, taxpayers, and governments alike. Patient data is usually spread across various data silos and databases that do not communicate well with one each other. This has led to gaps in the system where licensed pharmaceutical prescribers were unable to verify how many concurrent prescriptions a patient may have access to. Bitcoin has the potential to remove these data silos and improve public health through patient-controlled, auditable records. Example: **EHRData** [98]

- **Taxation** – Through Bitcoin and the power of microtransactions, government and business tax reporting become automated, audit friendly and extremely efficient. By integrating tax payment requirements into a programmable smart contract, payroll and other taxes become immediately available to the government allowing them secure payments faster, budget more accurately and decrease the risk for fraud.

- **Public Assistance** – Smart contracts on Bitcoin can also be used to create programmable tokens that could be utilized for government assistance programs such as the Food Stamp Program. These tokens can mitigate fraudulent use and prevent abuse through the ability to only approve the purchase specific needs-based items.

8. Blockchain and AI

The ancient myth of AI had developed through centuries: from the Greek, to the Age of Enlightenment [99], to the 20th Century when it made initial progress in the areas of game theory and theorem proofs. The modern concept of AI began to take shape in the 1950s after the arrival of the new computers made possible the design of reasoning processes that resembled those in human behavior. In this context, Alan Turing's 1950's "Turing Test" in *Computing Machinery and Intelligence* [100] provided a key step forward with a method for determining if a machine is "intelligent." Here, rather than asking whether the machine can think, the question changed to whether it can act as a thinker [101]. Seventy years later, AI tasks still struggle to reconcile the needs of sufficient representation, an effective and efficient decision-making mechanism that can make and execute timely decisions, and control.

Immutability, accessibility, non-repudiation, and decentralization of the data are some of the properties that allow blockchain technology to be used in AI developments, such as smart-contracts. Furthermore, the integration of the blockchain with AI provides solutions that can resolve problems intrinsic to the blockchain: for instance, by reducing energy consumption [102]. AI has also proven useful to better blockchain and smart contracts' security [103], for example, by helping in the process of code verification.

AI's technological capability to install cognitive capacities in machines so these can perform functions such as learn, interpret, and adapt, is related to consumed data. These data are often gathered from the users of smart telephones, and consumers of social media, and web applications [104]. As a result private and public organizations collecting these data, deal with issues of information centralization, legitimacy, authenticity, security, and privacy. Because data are centrally managed in AI projects, it can potentially be hacked and tampered with [105]. However, AI is also a tool that provides efficient solutions to major tasks such as in the allocation of resources, in managing large sets of data, and in procedural and repetitive tasks [106]. So the combined use of blockchain and AI addresses problems related to centralization, and offers solutions to issues related to the optimization of resources [107].

Intelligent and autonomous applications are designed to reduce human intervention in different types of processes; hence, their impact on individuals and societies raise important concerns. Harm to privacy, potential discrimination, limitation of citizenry choice and access to information, loss of skills, economic shocks, security of critical infrastructure, or long-term impacts on social well-being are just some of the key concerns these technological developments pose to society. That is the reason the development of these innovations need to be aligned with a set of defined values and ethical principles.

9. Ethical design framework

Given the ethical concerns these new technologies arise, a series of guidelines have been published by different institutions working at the crossroads of technology and social good. Here we will refer to those reported by the Beeck Center. [108] Nonetheless, others such as the *IEEE Global Initiative for Ethical Considerations in Artificial Intelligence and Autonomous Systems* [109], have also made huge efforts to encourage ethical considerations are prioritized when devising autonomous and intelligent technologies.

Establishing the ethical approach during the earliest phases of design is key when using Blockchain and AI. The reason is that changes will be more difficult to implement in later stages, if at all possible. This framework summarizes (p.21.):

> [] (1) give decision makers an outcome-focused and user-centric tool to assess the context-specific consequences and ethical implications of their blockchain design choices; and (2) to enable them to use this understanding to make the appropriate values-based design choices to achieve better social outcomes.

> [] Ultimately, these ethical considerations traced broadly to six root issues: governance, identity, access, verification and authentication, ownership of data, and security.

These factors are the basis for a three-phases framework. The first phase is a five-step process which establishes the intentionality of design with a focus on ethics. The second phase (p.40) is an iterative process which examines each design decision in light of the impacts it has on each other element of the ecosystem (i.e.: users, community...). The third phase (p.48), acknowledges that the context evolves in time and the relevance of each element changes. Hence, during this last phase there is a reevaluation of the first and second phases to assess significant changes in the environment.

Even though the implementation of such objectives will require additional time and resources dedicated at the start of each project, the benefits are self-evident even if just considering the impact on the smart contract environment. The reason is that smart contracts are deployed to start working when a predefined group of conditions are met. That is, the contracts will be triggered by inputs such as external events, information system sources, or other and these processes will be automatically enforced by algorithms unconstrained by ethical or legal considerations. Thus, in designing smart contracts, their impact beyond the realm of contract law should be analyzed. For instance, smart contracts could use ethically accepted rules when providing technological solutions and create models of governance through new social contracts. In this sense, the 2016 work of Reijers et al. [110]. analyzes how the modeling of blockchain governance reflects the key ideas of social contract theories. Their conclusions (p. 147–148) are that blockchain governance a) is justified by Rousseau's argument that it provides a solution to an existing structure of corrupted institutions; b) being non-discriminatory it reflects Rawls's "veil of ignorance," though power-relations are expressed in the public ledger; and c) acts in accordance to Hobbes idea of a *"totalitarian sovereign in terms of rule-enforcement, coupled with Rousseau's idea of decentralized governance and Rawls's idea of equal rights and liberties for all (that is, for all the nodes). Even though, it fails, to incorporate Rousseau's idea of the common good, and fails to implement conditions of distributive justice that Rawls thought to be essential for overcoming the initial situation."* (p.147).

Although the blockchain is perceived as a "neutral" technology, the political implications of its transformative power are profound as it will reconfigure economic, legal, institutional, and political spaces [111]. The information age promises great benefits from economies of scale and more efficient use of resources, but it also comes with a huge threatening potential to create masses of excluded individuals who cannot catch up with the times. Given the disconnect among different layers of citizens that it is likely to happen, renewed social contracts are essential to protect human dignity and the rights and opportunities of all [112].

Furthermore, any changes that make our democratic processes more transparent, inclusive, and participatory will benefit society. This was noted by Melanie Swan who in her 2015 work "Blockchain: Blueprint for a new economy, [113]" assured this technology will ease the appearance of new kinds of governance models and services. As an example, she mentions an increase of granular offer by which the government will design more targeted services. And she also enumerates a number of efforts to develop systems that will increase the quality of our democracies. For instance, she explains David Chaum's idea of random-sample elections [114]. Under this system, people selected randomly are asked to vote through an election website that contains candidate debates and activist sentiments. In David Chaum's view, because of cost reduction, many more consultative processes could be generated. Also, people would have time to inform themselves on whichever matter rather than be overwhelmed by political advertising. Furthermore, no government involvement would be necessary. A third idea discussed in Swan's book, is DAS which stands for distributed autonomous society. This model develops the principles for consensus-based decentralized governance systems and for decentralized voting systems. In her work, Professor Swan discusses this project as a form of delegated democracy, where voting power is vested in representatives. An example of such service is provided by https://liquidfeedback.org/, a company that offers an open source software to help present suggestions and make decisions. This is quite a compelling proposal because, under this method, people can align with each other on the bases of specific actions rather than "ideological" theories. Furthermore, power is not held long. Rather, individuals are responsible for a specific project. Thus, if standardized, this "liquid" in Liquid Democracy, would finish with political forms of permanent power as they are practiced today. Two immediate effects one can imagine would be a redistribution of power back to the people, and an increased impediment to the exercise of political corruption. Albeit there are many potential problems with this type of proposal, i.e.: power is obtained by groups which are already organized or citizens that might not wish to exercise these responsibilities, it might in fact provide a platform for a nation-wide discussion over the responsibilities of individuals on a modern technologically advanced society.

Overall, we can be sure that any elections properly organized using a voting protocol designed with blockchain and AI could be expected to exhibit at least the following desirable properties: privacy of the vote, perfect ballot secrecy, fairness, verifiability, self-tallying feature, dispute-freeness, fault tolerance, and resistance to serious failures. The works of Kiayias and Yung [115], Groth [116], Park et al. [117], Benaloh et al. [118], and Jonker et al. [119] provide a detailed description of these. We can also be sure that much upheaval would have been prevented if this would have been the underlying technology to the recent 2020 US Presidential elections.

10. Conclusion

The State should ensure the right of each individual to be secure in person and property and enhance the citizens' opportunities to make choices. Transparency and accountability are two key requirements to ensure the citizens' wills are not replaced by the needs of supra organizations: be it the state, large corporations, or the sole owners of certain resources. This is of particular importance in the age of "surveillance capitalism" when individuals might be looked upon and used as *"raw material supplies* [120]*".* It is in this environment that Bitcoin came to the market after both the 2008 white paper and the code were made available by Satoshi Nakamoto.

The 2009 birth of Bitcoin paved the way for a revolutionary transformation that announced the death of outdated technologies and evidenced the effort many across sectors and government, will have to make to say at par with the latest technology. This is a truly global solution that provides better transparency, fraud protection, it is faster, cheaper, and overall more efficient. Given this solution threatens to cause a fundamental and permanent change in our societies, and that the economic repercussions of the probable developments and trades are highly significant, public opinions have often been construed over a mixed of emotions and disinformation on the workings of the technology. As the Bitcoin builds untamperable public records in an efficient manner, a fear-mongering mentality intertwined with an problem posed by underlying conflict of interests has announced the "death" of this new sector repeatedly [121]. However, in just over a decade a myriad of transformative blockchain solutions have been built. Among the many, we have listed some ready solutions that will have immediate cathartic power. Of course these and other currently existing applications deserve a longer discussion.

During the decades following World War II, ethical standards were established to help govern how science in the future could move forward while not incurring the atrocities committed in the past [122]. Technology is considered as normatively neutral, but because transactions are irreversible and they solidify economic contracts by turning code into economic law, the use of Bitcoin poses a series of ethical questions. For instance, we could wonder about issues of privacy, whether miners are acting responsibly, whether this technology enables fraud, and so on. However, these questions can be answered by studying the technology itself and the trades. Here we turned our attention to whether the use of Bitcoin contributes to "ethics" according to the justice that is achieved when a society restores transparency and prevents fraud. In this imaginable future, Bitcoin will allow citizens have a more voluntary life and, in this way, it will contribute to the moral norm of justice by helping create a fairer society.

John Fitzgerald Kennedy [123] stated that *"change is the law of life. And those who look only to the past or present are certain to miss the future".* We hope this chapter contributes by helping the reader assess the depth of change this impending Bitcoin Revolution will unfold.

Bitcoin and Ethics in a Technological Society
DOI: http://dx.doi.org/10.5772/intechopen.96798

Author details

Eva R. Porras[1*] and Bryan Daugherty[2]

1 Department of Business Economics, Applied Economics II, and Fundamentals of Economic Analysis at Universidad Rey Juan Carlos, Madrid, Spain

2 Bitcoin Association, Zug, Switzerland

*Address all correspondence to: eva_porras@hotmail.com

IntechOpen

References

[1] Please note that Bitcoin refers to the protocol, while bitcoin refers to the currency itself.

[2] please see https://ramonquesada. com/ and https://es.ramonquesada.com/ glossary/bitcoin-es/; https://www.mail- archive.com/cryptography@metzdowd. com/msg10142.html

[3] https://bitcoin.org/bitcoin.pdf Satoshi Nakamoto Bitcoin: A Peer-to- Peer Electronic Cash System

[4] Throughout this text we use cryptocurrency, digital money and digital currencies as synonyms. These terms refer to native digital currencies, such as bitcoin, and not to the digital version of fiat monies, such as the euro, dollar, pound, etc. We use the term "Bitcoin" to refer to the overall Bitcoin network, while "bitcoin" refers to digital money produced by this network.

[5] Paul Krugman's article "Bitcoin Is Evil" December 28, 2013 https://krugman.blogs.nytimes. com/2013/12/28/bitcoin-is-evil/?_r=1

[6] Paul Krugman's article "Bitcoin Is Evil" December 28, 2013 https://krugman.blogs.nytimes. com/2013/12/28/bitcoin-is-evil/?_r=1

[7] Charlie Stross article "Why I want Bitcoin to die in a fire" December 28ck, 2013 https://www.antipope.org/ charlie/blog-static/2013/12/why-i-want- bitcoin-to-die-in-a.html

[8] William Suberg's article "JPMorgan CEO Jamie Dimon Returns to Bitcoin Bashing, Calls Cryptocurrency a 'Scam' "AUG 06, 2018 https://cointelegraph. com/news/jpmorgan-ceo-jamie-dimon- returns-to-bitcoin-bashing-calls- cryptocurrency-a-scam

[9] https://cointelegraph.com/news/ jp-morgan-chase-ceo-jamie-dimon- bitcoin-is-going-to-be-stopped

[10] http://www.smh.com.au/ technology/technology-news/drugs- bought-with-virtual-cash-20110611- 1fy0a.html; https://upload.wikimedia. org/wikipedia/commons/0/0b/ Silk_Road_Seized.jpg

[11] https://www.activism.net/ cypherpunk/crypto-anarchy.html

[12] https://www.activism.net/ cypherpunk/manifesto.html

[13] W. Dai, "b-money," http://www. weidai.com/bmoney.txt, 1998. http:// www.weidai.com/bmoney.txt

[14] https://www.Cypherpunks.to/list/

[15] David Chaum, "Achieving Electronic Privacy," *Scientific American*, August 1992, 96-101.

[16] David Chaum, "Untraceable Electronic Mail, Return Addresses, and Digital Pseudonyms" Communications of the ACM. Volume 24 Issue 2, Feb. 1981, Pages 84-90 http://citeseerx.ist. psu.edu/viewdoc/download?doi=10.1.1. 79.7468&rep=rep1&type=pdf

[17] David Chaum, "Blind signature for untraceable payments" *Advances in Cryptology Proceedings of Crypto. 82 (3): 199-203.* doi:10.1007/978- 1-4757-0602-4_18. (http://www. hit.bme.hu/~buttyan/courses/ BMEVIHIM219/2009/Chaum. BlindSigForPayment.1982.PDF

[18] http://nakamotoinstitute.org/ finney/rpow/

[19] Fineman, S. (1993). Emotion and organizing. In C. Hardy (Ed.), *Handbook of organization studies* (pp. 1-64). London: Sage.

[20] https://cointelegraph.com/news/94- year-old-berkshire-hathaway-vp-its- disgusting-people-buy-bitcoin

[21] https://www.cnbc.com/2018/03/05/
bitcoin-more-likely-to-be-100-
than-100000-in-10-years-kenneth-
rogoff.html

[22] William Suberg's article "JPMorgan
CEO Jamie Dimon Returns to Bitcoin
Bashing, Calls Cryptocurrency a 'Scam'"
Aug 06, 2018 https://cointelegraph.
com/news/jpmorgan-ceo-jamie-dimon-
returns-to-bitcoin-bashing-calls-
cryptocurrency-a-scam

[23] https://cointelegraph.com/news/
jp-morgan-chase-ceo-jamie-dimon-
bitcoin-is-going-to-be-stopped

[24] https://www.theguardian.com/
technology/2017/sep/13/bitcoin-fraud-
jp-morgan-cryptocurrency-drug-dealers

[25] https://twitter.com/CNBC/
status/907663994613354496

[26] https://cointelegraph.com/news/
bitcoin-mining-thrives-in-venezuela-
thanks-to-hyperinflation-and-free-
electricity

[27] https://cointelegraph.com/news/
bitcoin-in-venezuela-bolivar-worth-50-
of-world-of-warcraft-gold

[28] https://www.theguardian.com/
technology/2017/sep/13/bitcoin-fraud-
jp-morgan-cryptocurrency-drug-dealers

[29] https://www.theguardian.com/
technology/2017/sep/13/bitcoin-fraud-
jp-morgan-cryptocurrency-drug-dealers

[30] https://cointelegraph.com/
news/jpmorgan-ceo-jamie-dimon-
returns-to-bitcoin-bashing-calls-
cryptocurrency-a-scam

[31] William Suberg Bitcoin Price Drops
After JPMorgan CEO Harsh Rhetoric
SEP 13, 2017 https://cointelegraph.
com/news/bitcoin-price-drops-after-
jpmorgan-ceo-harsh-rhetoric

[32] https://www.theguardian.
com/technology/2017/sep/13/

bitcoin-fraud-jp-morgan-
cryptocurrency-drug-dealers

[33] Paul Krugman's article "Bitcoin
Is Evil" December 28, 2013
https://krugman.blogs.nytimes.
com/2013/12/28/bitcoin-is-evil/?_r=1

[34] Charlie Stross article "Why I want
Bitcoin to die in a fire" December
28ck, 2013 https://www.antipope.org/
charlie/blog-static/2013/12/why-i-want-
bitcoin-to-die-in-a.html

[35] https://www.theguardian.
com/technology/2018/may/21/
cryptocurrencies-fad-money-bitcoin

[36] https://www.theguardian.
com/technology/2018/may/21/
cryptocurrencies-fad-money-bitcoin In
a recent article by 2013 Nobel Laureate
Robert Shiller, the economist points
out that

[37] Jenkinson, G. (1918) Tulips,
Bubbles, Obituaries: Peering Through
the FUD About Crypto. https://
cointelegraph.com/news/tulips-
bubbles-obituaries-peering-through-
the-fud-about-crypto. Published Jun
24, 2018.

[38] https://99bitcoins.com/bitcoin-
obituaries/ collected by 99bitcoins.com

[39] https://cointelegraph.com/
tags/altcoin

[40] https://digiconomist.
net/?s=Ethereum+obituaries;
https://digiconomist.net/
new-ethereum-obituaries

[41] Following Krugman's article there
are 524 comments, Charlie Stross'
Comments (902)

[42] https://cointelegraph.com/news/
volatility-the-necessary-evil-of-
cryptocurrency-and-how-to-handle-it

[43] https://twitter.com/ErikVoorhees

[44] https://en.wikipedia.org/wiki/
Troubled_Asset_Relief_Program

[45] https://cointelegraph.com/news/
jp-morgan-sees-crypto-as-competition-
and-risk-to-its-business-in-sec-
annual-report

[46] Molly Jane Zuckerman J.P Morgan
Sees Crypto As 'Competition' And 'Risk'
To Its Business In SEC Annual Report
FEB 28, 2018 https://cointelegraph.
com/news/jp-morgan-sees-crypto-as-
competition-and-risk-to-its-business-
in-sec-annual-report

[47] http://investor.shareholder.
com/jpmorganchase/secfiling.
cfm?filingID=19617-18-
57&CIK=19617#CORP10K2017_HTM_
S440D20F00AA0567AA
DC9B36846A275C5

[48] https://cointelegraph.com/tags/
bitcoin-futures

[49] https://cointelegraph.com/news/
elite-investment-bank-goldman-sachs-
to-clear-bitcoin-futures-for-clients

[50] https://www.bloomberg.com/
news/articles/2017-12-21/goldman-is-
said-to-be-building-a-cryptocurrency-
trading-desk

[51] https://www.nytimes.
com/2018/05/02/technology/bitcoin-
goldman-sachs.html

[52] https://cointelegraph.com/news/
bloomberg-goldman-sachs-to-setup-
cryptocurrency-trading-desk

[53] https://www.sec.
gov/Archives/edgar/
data/70858/000007085818000009/bac-
1231201710xk.htm

[54] https://cointelegraph.com/news/
bank-of-america-our-inability-to-
adapt-could-see-a-failure-to-compete-
with-crypto

[55] https://cointelegraph.com/news/
bank-of-america-our-inability-to-

adapt-could-see-a-failure-to-compete-
with-crypto

[56] https://cointelegraph.com/news/
jp-morgan-chase-bans-buying-
cryptocurrency-with-credit-cards

[57] https://www.sec.
gov/Archives/edgar/
data/70858/000007085818000009/bac-
1231201710xk.htm

[58] https://cointelegraph.com/news/
crypto-unlikely-to-disappear-says-
internal-report-attributed-to-jp-morgan

[59] Krugman, P. R. (2013). *Bitcoin is
evil*. Retrieved from http://krugman.
blogs.nytimes.com/2013/12/28/
bitcoin-is-evil/?_r=1

[60] https://twitter.com/CNBC/
status/907663994613354496

[61] https://robertskidelsky.
com/2018/05/23/
why-reinvent-the-monetary-wheel/

[62] https://youtu.be/ERImeMPRFRE.
3. Donald Trump, the Cryptography
Mailing List, and SIGHASH types -
Satoshi Nakamoto - CSW & RXC Posted
online on January 21st, 2021 minutes
00:49-1.21

[63] https://www.govinfo.gov/content/
pkg/PLAW-105publ304/pdf/PLAW-
105publ304.pdf Digital Millennium
Copyright Act Signed into law by
President Bill Clinton on October
28, 1998.

[64] Dr Craig Wright. 11 Jan 2021. An
exploration of ingroup behaviour
and social psychology in developing
socially abhorrent behaviours in social
media and financial systems. Abstract.
https://papers.ssrn.com/sol3/papers.
cfm?abstract_id=3763746

[65] Dr Craig Wright. 11 Jan 2021. An
exploration of ingroup behaviour
and social psychology in developing

socially abhorrent behaviours in social media and financial systems. Abstract. https://papers.ssrn.com/sol3/papers.cfm?abstract_id=3763746 pages 9 and 10.

[66] p322-331 R. Z. SAGDEEV] *Science and Revolutions*, The Tanner Lectures on Human Value. March 9-16, 1992. https://tannerlectures.utah.edu/_documents/a-to-z/s/Sagdeev93.pdf

[67] p322-331 R. Z. SAGDEEV] *Science and Revolutions*, The Tanner Lectures on Human Value. March 9-16, 1992. https://tannerlectures.utah.edu/_documents/a-to-z/s/Sagdeev93.pdf

[68] f p322-331 R. Z. SAGDEEV] *Science and Revolutions*, The Tanner Lectures on Human Value. March 9-16, 1992. https://tannerlectures.utah.edu/_documents/a-to-z/s/Sagdeev93.pdf

[69] p322-331 R. Z. SAGDEEV] *Science and Revolutions*, The Tanner Lectures on Human Value. March 9-16, 1992. https://tannerlectures.utah.edu/_documents/a-to-z/s/Sagdeev93.pdf

[70] https://www.washingtontimes.com/topics/parler/ ; https://www.washingtontimes.com/news/2021/jan/24/parler-provided-freedom-censorship-cost-users-data/

[71] 2013 Nobel Laureate Robert Shiller said when referring to the case of digital cash: *"statement of faith in a new community of entrepreneurial cosmopolitans who hold themselves above national governments"*

[72] https://www.lainformacion.com/mercados-y-bolsas/la-cnmv-y-el-banco-de-espana-alertan-del-riesgo-regulatorio-sobre-bitcoin/2828815/ https://www.lainformacion.com/economia-negocios-y-finanzas/economia-criptomonedas-no-tienen-garantias-otros-activos/2828975/https://www.expansion.com/mercados/div isas/2021/02/10/6023b9bfe5fdea557 78b4619.html

[73] Jürgen Habermas1986, Lecture One: How Is Legitimacy Possible On The Basis Of Legality?

[74] Max Weber, Wirtschaft und Gesellschaft (Cologne, 1964), ch. 3, pp. 2, 160ff.

[75] https://www.youtube.com/watch?v=LdvQTwjVmrE&feature=emb_logo All-Star Panel: Ed Moy, Joseph VaughnPerling, Trace Mayer, Nick Szabo, Dr. Craig Wright. Presented at Bitcoin Investor Conference - Las Vegas, NV Oct. 29-30, 2015 http://BitcoinInvestor.com

[76] https://www.youtube.com/watch?v=LdvQTwjVmrE&feature=emb_logo See minutes (48:18-50:01)

[77] https://www.youtube.com/watch?v=LdvQTwjVmrE&feature=emb_logo All-Star Panel: Ed Moy, Joseph VaughnPerling, Trace Mayer, Nick Szabo, Dr. Craig Wright. Presented at Bitcoin Investor Conference - Las Vegas, NV Oct. 29-30, 2015 http://BitcoinInvestor.comsee minutes 52:22-53:42

[78] https://www.youtube.com/watch?v=LdvQTwjVmrE&feature=emb_logo All-Star Panel: Ed Moy, Joseph VaughnPerling, Trace Mayer, Nick Szabo, Dr. Craig Wright. Presented at Bitcoin Investor Conference - Las Vegas, NV Oct. 29-30, 2015 http://BitcoinInvestor.comsee minutes 63:25-63:58

[79] Baumol, W.J. (2002). The free-market innovation machine: Analyzing the growth miracle of capitalism. Princeton university press, in J.M. Rivière, Blockchain Technology And IP - Investigating Benefits And Acceptance In Governments And Legislations. *Junior Management Science,* 2018, 3(1), 1-15.

[80] Tapscott, D., Tapscott, A. (2016). *Blockchain Revolution: How the Technology Behind Bitcoin is Changing*

Money, Business, and the World. Penguin Canada.

[81] Tapscott, D., Tapscott, A. (2016). The impact of the blockchain goes beyond financial services. *Harvard Business Review*. [Accessed Dec.27, 2020] https://hbr.org/2016/05/the-impact-of-the-blockchain-goes-beyond-financial-services?autocomplete=true.

[82] James Madison, Letter to W. T. Barry, August 4, 1822. "From James Madison to William T. Barry, 4 August 1822," *Founders Online,* National Archives, https://founders.archives.gov/documents/Madison/04-02-02-0480. [Original source: *The Papers of James Madison*, Retirement Series, vol. 2, *1 February 1820–26 February 1823*, ed. David B. Mattern, J. C. A. Stagg, Mary Parke Johnson, and Anne Mandeville Colony. Charlottesville: University of Virginia Press, 2013, pp. 555-558.]

[83] https://www.pewresearch.org/politics/2019/04/11/public-trust-in-government-1958-2019/

[84] https://ycnews.com/political-polarization-continues-to-consume-the-vitality-of-the-united-states/

[85] https://www.washingtontimes.com/news/2021/jan/21/leftists-portland-protest-biden-inauguration-vanda/

[86] https://obamawhitehouse.archives.gov/the-press-office/2011/04/27/executive-order-13571-streamlining-service-delivery-and-improving-custom

[87] Cyber-Security and Infrastructure Agency revealed that a yearlong hack affected US government agencies and critical infrastructure entities

[88] American businesses and taxpayers over $100 billion dollars. https://www.rollcall.com/2021/01/11/cleaning-up-solarwinds-hack-may-cost-as-much-as-100-billion/

[89] https://papers.ssrn.com/sol3/papers.cfm?abstract_id=3440802 Bitcoin: A Peer-toPeer Electronic Cash System Dr Craig S Wright.

[90] https://www.govloop.com/community/blog/built-to-fail-why-governments-struggle-to-implement-new-technology/

[91] https://tokenized.com/

[92] https://nchain.com/

[93] https://legallychained.com/

[94] https://www.elas.digital/

[95] https://bitcoinassociation.net/ballots-on-blockchain/

[96] https://unisot.com/

[97] http://www.vxpass.com/

[98] https://ehrdata.com

[99] https://www.marxists.org/reference/archive/la-mettrie/1748/man-machine.htm

[100] https://www.csee.umbc.edu/courses/471/papers/turing.pdf

[101] https://www.historyofinformation.com/detail.php?id=4289

[102] J. Chen, K. Duan, R. Zhang, L. Zeng, and W. Wang, "An AI Based Super Nodes Selection Algorithm in BlockChain Networks." [Online]. Available: https://arxiv.org/pdf/1808.00216.pdf

[103] T. Marwala and B. Xing, "Blockchain and Artificial Intelligence," *arXiv preprint arXiv:1802.04451*, 2018. [Online]. Available: http://arxiv.org/abs/1802.04451.

[104] K. Salah, M. H. U. Rehman, N. Nizamuddin and A. Al-Fuqaha, "Blockchain for AI: Review and Open

Research Challenges," in IEEE Access, vol. 7, pp. 10127-10149, 2019. doi: 10.1109/ACCESS.2018.2890507

[105] Diversifying Data With Artificial Intelligence And Blockchain Technology, 2018, retrieved from https://www.forbes.com/sites/rachelwolfson/2018/11/20/diversifying-data-with-artificial-intelligence-and-blockchain-technology/#71272a104dad

[106] A. Androutsopoulou, N. Karacapilidis, E. Loukis, Y. Charalabidis, Transforming the communication between citizens and government through AI-guided chatbots, Government Information Quarterly, Volume 36, Issue 2, 2019, Pages 358-367, ISSN 0740-624X, https://doi.org/10.1016/j.giq.2018.10.001

[107] Gabriel Axel Montes, Ben Goertzel, Distributed, decentralized, and democratized artificial intelligence, Technological Forecasting and Social Change, Volume 141, 2019, Pages 354-358, ISSN 0040-1625, https://doi.org/10.1016/j.techfore.2018.11.010

[108] 2019. Cara Lapointe and Lara Fishbane. The-Blockchain-Ethical-Design-Framework.pdf https://www.researchgate.net/publication/330069634_The_Blockchain_Ethical_Design_Framework #fullTextFileContent

[109] Ethically aligned design IEEE A vision for Prioritizing Human Wellbeign with Artificial Intelligence Autonomous Systems ead_v1.pdf https://standards.ieee.org/content/dam/ieee-standards/standards/web/documents/other/ead_v1.pdf

[110] Reijers, W., O'Brolcháin, F. and Haynes, P. Governance in Blockchain Technologies & Social Contract Theories. *Ledger*, 1, 0 (2016), 134-151. https://www.researchgate.net/publication/312244646_Governance_in_Blockchain_Technologies_Social_Contract_Theories

[111] Coeckelbergh, M., Reijers, W. "Crypto currencies as narrative technologies." *SIGCAS Computers & Society*, 45.3 172-178 (2015) https://www.researchgate.net/publication/281586152_Cryptocurrencies_as_Narrative_Technologies

[112] Mason, R. O. Four Ethical Issues of the Information Age. *MIS Quarterly*, 10, 1 (March 1986), 5-12. file:///C:/Users/Usuario/AppData/Local/Temp/mason.pdf

[113] Melanie Swan (2015) "Blockchain: Blueprint for a new economy" Published by O'Reilly Media, Inc., 1005 Gravenstein Highway North, Sebastopol, CA 95472.

[114] Chaum, D. "Random-Sample Elections: Far Lower Cost, Better Quality and More Democratic." Accessed 2021. https://www.chaum.com/publications/ Random-Sample%20Elections.pdf.

[115] Aggelos Kiayias and Moti Yung. 2002. Self-tallying Elections and Perfect Ballot Secrecy. In Public Key Cryptography, David Naccache and Pascal Paillier (Eds.). Springer Berlin Heidelberg, Berlin, Heidelberg, 141-158.

[116] Jens Groth. 2004. Efficient Maximal Privacy in Boardroom Voting and Anonymous Broadcast. In Financial Cryptography, Ari Juels (Ed.). Springer Berlin Heidelberg, Berlin, Heidelberg, 90-104.

[117] Sunoo Park, Michael Specter, Neha Narula, and Ronald L Rivest. 2020. Going from bad to worse: from internet voting to blockchain voting. https://people.csail.mit.edu/rivest/pubs/PSNR20.pdf

[118] Josh Benaloh, Ronald Rivest, Peter YA Ryan, Philip Stark, Vanessa Teague, and Poorvi Vora. End-to-end verifiability.

arXiv preprint arXiv:1504.03778
(2015). https://arxiv.org/ftp/arxiv/
papers/1504/1504.03778.pdf

[119] Hugo Jonker, Sjouke Mauw, and Jun
Pang. 2013. Privacy and verifiability in
voting systems: Methods, developments
and trends. Computer Science Review
10 (2013), 1-30.

[120] Zuboff, 2019: 87. *The Age of
Surveillance Capitalism: The Fight for
a Human Future at the New Frontier
of Power*

[121] https://cointelegraph.com/news/
tulips-bubbles-obituaries-peering-
through-the-fud-about-crypto

[122] Dorothy Roberts - *The Old Biosocial
and the Legacy of Unethical Science*
103 p.346

[123] *Kennedy, J.F. (1963). Public Papers
of the Presidents.* [Accessed Jan.25,
2021] https://www.jfklibrary.org/learn/
about-jfk/life-of-john-f-kennedy/
john-f-kennedy-quotations

Chapter 3

How Blockchain and AI Enable Personal Data Privacy and Support Cybersecurity

Stanton Heister and Kristi Yuthas

Abstract

Recent increases in security breaches and digital surveillance highlight the need for improved privacy and security, particularly over users' personal data. Advances in cybersecurity and new legislation promise to improve data protection. Blockchain and distributed ledger technologies provide novel opportunities for protecting user data through decentralized identity and other privacy mechanisms. These systems can allow users greater sovereignty through tools that enable them to own and control their own data. Artificial intelligence provides further possibilities for enhancing system and user security, enriching data sets, and supporting improved analytical models.

Keywords: personally-identifiable data (PII), personal-data privacy, decentralized identity (DID), self-sovereign identity, cybersecurity, GDPR, zero-knowledge proofs

1. Introduction

The amount of personal data being collected is rapidly proliferating. Enterprises and governments use this data to profile individuals and to predict and control their attitudes and behavior. This can result in customized experiences, personalized services, and more efficient use of resources. It can also result in misinformation and exploitation by the entity that collected the data or by others that purchase or steal it. In response to increases in cybercrime and growing consumer concern, legislation to protect personal data is being proposed and implemented. Organizations trading in personal data face increasing costs associated with managing and securing data. They also face increasing risks that data will be misused or stolen, and that they will face legal or financial consequences, as well as damage to both their reputation and to relationships with customers and other stakeholders.

In this chapter, we explore how blockchain and artificial intelligence can offer solutions for protecting and securing personal data. Decentralized and federated identify systems provide users control over what, when and how much of their personal information can be shared and with whom. These systems can also reduce cybersecurity threats. Artificial intelligence complements blockchain-based privacy solutions by enabling users to better manage their data and by ensuring that data and models derived from the data are more accurate, fair, and reliable.

2. Personal data privacy

A foundational privacy issue facing information system developers and users is personal data privacy. Personally-identifiable data about clients, employees, prospects and other stakeholders may be regularly collected and stored in shared ledgers. Today, many organizations store private stakeholder data and even passwords in unencrypted form. Even when data are encrypted or anonymized, it may be possible to identify users unless well-developed cybersecurity processes are designed into data management systems. With frequent cybersecurity failures and increasing regulation, maintaining the privacy of personally identifiable information (PII) has become an issue of strategic concern for many organizations.

PII includes any data that can be traced back to a specific person, and can include individual items such as biometric data, social security numbers, phone numbers, or geolocation data. PII can also include data combinations, such as postal codes, birthdates, and gender, or behavioral data associated with one person. Organizations gather and store personal data about current and future customers and employees as well as about other stakeholders.

3. Cybersecurity and privacy breaches

Cybersecurity has become increasingly important for governments and businesses alike. Information security—one component of cybersecurity—focuses on protecting the integrity and privacy of data as it is captured, stored and used. The people, processes, and technology associated with data work in concert to create and maintain security.

Despite advances in security protocols and software, privacy breaches are on the rise. According to Risk Based Security's 2020 data breach report, "The total number of records compromised in 2020 exceeded 37 billion, a 141% increase compared to 2019" [1]. Personal records of system users are regularly compromised, and millions of these records, including names, emails and passwords, have been subject to data breaches, in many cases even including addresses, birth dates and financial information [1].

A data breach occurs from unauthorized access to an organization's database, enabling cyber hackers to steal sensitive personal information such as passwords, credit card numbers, social security numbers, and banking information [2]. These well documented breeches have had adverse consequences, including credit card fraud, and identity theft, which can have lasting negative effects on personal credit, often taking months, if not years, to remedy [2]. Some of the Largest, most recent cyber hacks include the 2013/14 breech of Yahoo's database by what is thought to have been a state-sponsored cyberattack, impacting over 3 billion users. The hackers collected consumers' names, email addresses, telephone numbers, dates of birth, hashed passwords and unencrypted answers to security questions.

In 2017, the credit reporting agency Equifax was subject to a cyberattack in which affected an estimated 143 million consumers. System administrators weren't aware of the suspicious activity for two months and did not report the breach for a full month after its discovery. It is believed that Equifax was breached by Chinese state-sponsored hackers engaged in espionage [3]. The collective financial impact to individual victims is not known, nor is it known what security and strategic damage was incurred by the state, but these cases highlight the potential risk when PII are housed in a centralized data base.

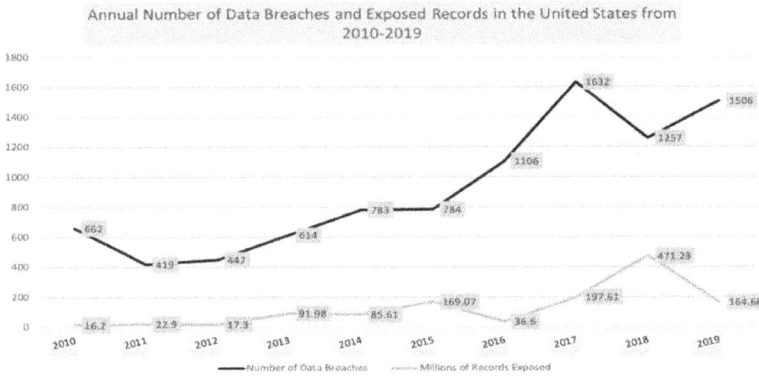

Figure 1.
Cybersecurity breaches and record exposure.

Most of the data gathered and stored are in the control of governments and corporations, which have gathered volumes of personal information that they are responsible for securing. At the same time, these organizations may be monetizing these datasets, either by using them to improve their own operations and offerings or by selling them to third parties. The volume of data generated and collected is increasing exponentially, enlarging the footprints of users. Data consolidators are able to link data elements across data sources and combine data in ways that were never anticipated by the parties that collected the information nor by the users that provided it.

Figure 1, which uses from data provided by Statista [4], shows the cost of amassing these large databases. Statista, a statistical research firm, tracks cybersecurity failures and trends. A recently published Statista report reveals that these events are increasing, especially in the past five years, underscoring the need to improve how data are secured. It should be noted that in 2020 a massive cyber breach by what is thought to be Russia could result in higher numbers for 2020 especially in the records exposed category as it is thought to be significant. The extent of the breach is still under investigation at the time of this publication.

4. Privacy regulations

The right to privacy is a considered to be basic human right in many parts of the world. That privacy may extend to individuals' right to control their own personal data. This right must be carefully defended as ownership and management of an individual's personal data can impact relationships with others and even the data-owner's identity [5].

Regulations governing how personal data are gathered and managed are rapidly being developed. The European Union has led the way in legislating privacy law through the General Data Protection Regulation (GDPR), passed in 2016. The law requires organizations, that gather personal data about EU citizens for transactions with EU member states, must carefully protect that data to ensure privacy.

In the US, the California Privacy Rights Act (CPRA) which expands on the 2018 California Consumer Privacy Act (CCPA), adopts many principles from the GDPR [6]. The CCPR is designed to provide residents of California the right:

1. to know what personal data is being collected

2. to know whether it is being sold or disclosed and to whom

3. to refuse the sale of their personal data

4. to access their personal data

5. to request that a business delete any personal data

6. not to be discriminated against for exercising their privacy rights [7]

At the federal level, the Consumer Online Privacy Rights Act (COPRA) was introduced in December 2019 by Democratic senators, led by Maria Cantwell. Although this bill has yet to pass, and previous federal privacy bills have failed, governmental bodies continue to pursue stricter laws for governing data [8].

Privacy laws directly affect how companies operate and will require firms that use consumer data to implement systems and operational practices that enable them to conform to these new regulations. Blockchain and Distributed Ledger Technology are uniquely positioned to help companies comply with existing and potential future regulation as it relates to personal property and data privacy.

5. Blockchain and privacy

Among the significant benefits of blockchain solutions is that they enable organizations to share data in ways not previously available, opening up possibilities for enhanced collaboration, improved operational efficiencies and expanded revenue. Questions about how to maintain privacy over the data are heightened in these environments because the data are stored in shared ledgers which may be accessible by multiple blockchain participants.

ConsenSys, a blockchain technology solutions company, in discussing the security of public blockchains, argues that "In reality, privacy is not a property of any blockchain. Rather, there are layers of privacy that can be applied to any blockchain..." [9]. Designers must carefully consider which parties are allowed to read and write transactions and how transactions are broadcast, validated, and stored. Additional issues relating to how permissions and security measures are updated and enforced are also important considerations. Decisions about who owns the data and how data can be used by organizations and computer applications further complicates privacy discussions [9].

5.1 Decentralized identity

Self-sovereign identity, a widely held view among blockchain proponents, holds that individuals should have control over their own identities and should have autonomy over how facets of identity are shared with others. Decentralized identity (DID) is a blockchain-enabled embodiment of self-sovereign identity that can profoundly improve the privacy and security of personal data.

DID refers to individual ownership of personal digital data relating to many elements of identity. Microsoft, which participates in defining DID standards, takes the perspective of the individual. "Currently, our identity and all our digital interactions are owned and controlled by other parties, some of whom we aren't even aware of [10]." Returning ownership of data to the individuals to whom the data pertains can provide benefits both to those individuals and to organizations that would otherwise be responsible for protecting the data.

Blockchain technology enables DID and provides a way for individuals to store their own data outside of the databases of the parties with whom they transact. Data are owned and controlled by these individuals and pointers to this data or metadata can be stored on the blockchain and can be used to verify the validity of claims the users make about their personal data. For example, a driver's license bureau might issue a driver's license to a user, which the user stores privately. When an insurance company or other party wishes to verify that the user is licensed, the user can present the license to a party such as an insurance company, and the party can independently verify the issuer and expiration date.

Anyone can create a DID. When this identity is first created, there is no information attached to it. Over time, the user could attach a driver's license or other identifying data to that DID. The process that a third party might use to verify that a particular person owns a DID, is similar to the process of validating that a person owns an email address. For example, an online gaming account can be attached to an email address. A party seeking to validate that a person was the owner of that account could send a private message, such as a security code, to the email address and ask the person to provide that code, something that only the person possessing the password for that email address could provide.

Unlike an email account, the DID would be owned and stored by a person rather than by an email service provider. The password, or private key, would also be secured by the owner. Personal information relating to the identity could be stored in an identity hub—an encrypted repository of personal data that is stored outside the blockchain, likely in a combination of phone, PC, and cloud data or offline storage devices [10]. Through the use of an identity hub, the person could control which pieces of information to share with an external party.

DIDs reduce the probability of unwanted correlation. The use of common identifiers—such as email addresses on different web sites—creates what is called a correlation problem. Correlation in this context means entities can, without a user's consent, associate information about a single identity across multiple systems. Email addresses utilize data on almost every website. When users provide the same email address on different sites—along with perhaps additional pieces of personal information like a phone number or physical address—they unknowingly enable a potential for correlation. In this case, entities can correlate that data across sites.

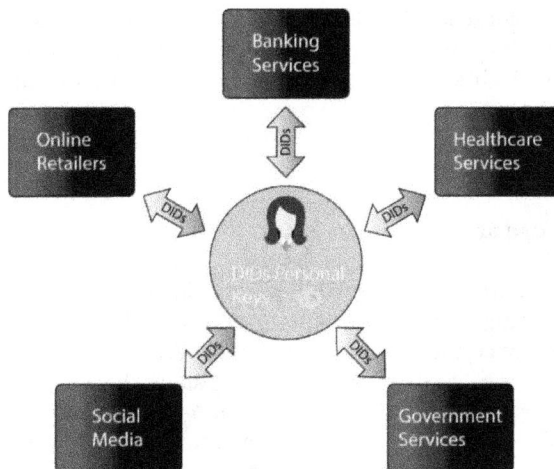

Figure 2.
Decentralized identities and service providers.

Tracking cookies and web clicks enable the linking of IDs across websites which can result in outsiders gaining a full picture of users' identity, where they live, their gender, age range, interests, and other information [10].

Figure 2 depicts how a user of several services and on-line websites can store data in a central user-controlled location and interact separately with each service provider. This enables the user to control the specific pieces of information that can be seen by each provider.

5.2 Blockchain-enabled federated identity

DIDs can help users secure and control their data property and determine who gets access to that data. Blockchains can also increase security for individuals when interacting with multiple internet platforms or services through the use of decentralized federated identities.

Blockchains allow entities to protect privacy of individuals—central to self-sovereign identity. Traditionally, users of a system or set of systems possess what is referred to as a federated identity, which can be described as a single identity used by individuals to access services or information platforms, provided by multiple parties, whereby a single identity is enabled and determined by single sign on (SSO) authentication. Consider a health care network that includes multiple entities like hospitals, insurance carriers, or urgent care clinics, where the providers enable the use of a single sign-on credential or *digital federated identity* to access all services. This type of identity, which is typically stored and managed in a central location by a service provider, is prone to security vulnerabilities [11].

The distributed nature of Blockchain technology provides an opportunity for networks to enable single sign-on, or federated identities much more securely. ElGayyar [11] proposes a blockchain-based federated identity framework (BFID) where the network of providers themselves, rather than a centralized third party, manage the system, identification, and authentication of the users. Any entity within the blockchain network can verify credentials and issue the identity for any user in the system. In a BFID, all transactions are written and maintained within the blockchain where the system takes advantage of the secure and immutable nature of the distributed ledger, thereby practically eliminating the possibility for identity breaches and potential theft.

Blockchain-based federated identity frameworks can be configured on both public and private blockchain implementations and make use of smart contracts to react to potential rule changes that may occur while governing identity management within the system. Additionally, these frameworks enable users to audit and control how their identities are used while also providing the network business entities the ability to monitor how their services are being used, enabling process improvement and a better overall user experience.

5.3 Zero-knowledge proofs

Zero-knowledge proofs enable ease of access to identity and other important data while maintaining privacy and property control for individuals. Zero-knowledge proofs are cryptographic methods whereby a user or "prover" can convince someone, or a "verifier" that something about them is true without providing, revealing or sharing that information. A common example is a customer attempting to order an alcoholic beverage from a bartender who demands to know that the patron is 21 of age or older. Providing a driver's license reveals the patron's full birth date as well as height, eye color, and home address—information that could be misused or stolen.

Zero-knowledge proofs use cryptographic algorithms that enable a prover to mathematically demonstrate to a verifier that a statement is correct without revealing any data. When the state issues a 21-and-over driver's license, it asks the driver to type in a secret nickname, unknown to the licensing bureau. This nickname could then be hashed together with the driver's license number, and stored in a public list representing valid drivers over 21. At the bar, you could type your nickname and license number into a hash generator, and if the resulting hash matched one on the list, the bartender would know that you were of legal age [12].

There are two types of zero knowledge proofs, interactive and non-interactive. Most commonly, zero knowledge protocols are interactive whereby the prover (an individual or more likely a computer) and the verifier participate in a back and forth set of questions or challenges that, when answered correctly a given number of times, enables the prover to convince the verifier, with very high probability, that the statement they are making is true.

An example of an interactive zero knowledge proof could involve two colored balls that are identical in every way accept their color. One is red and one green. Let's assume the verifier is completely color blind and cannot tell the color of either ball. You want to prove to the verifier that the balls do in fact differ in color. The verifier puts the balls behind their back and shows one. The prover indicates the color. The verifier then does this again and asks if they switched the ball. Since you can see the different colors you can say with certainty that the ball was either switched or not. After several rounds of this, it becomes more statistically true that there are in fact two different balls as the probability that you could guess correctly over and over goes down to almost zero [13].

Non-interactive proof is more like the example above of the patron proving their age to a bartender with a proof statement that reveals age but not additional information that might be revealed if the prover were to show their photo. Proving which point Value a card in a deck of 52 cards, without identifying its suit, can provide an example of this type of proof. The prover states that the card they are holding is a king but does not want to reveal which king—the king of hearts, diamonds, spades or clubs. If the cryptographic string also contains information that reveals the other 48 cards, none of which are kings, we can know for certain that the prover does in fact hold a king of some kind.

Zero knowledge proofs are powerful tools for maintaining privacy and property control for individuals that may need to provide a bit of personal information but no more than absolutely necessary.

6. Artificial intelligence and privacy

Artificial Intelligence (AI) is a broad field that includes machine learning and cognitive computing where computers are programed to mimic human cognitive functions such as learning and problem solving but many times, much faster and in more accurate ways [14]. The use of AI is expanding into a plethora of areas including speech recognition, facial recognition, medical diagnosis, financial predictions, tracking of disease outbreaks etc. AI algorithms enable computing systems to rationalize and take actions aimed at achieving a specific goal or set of goals.

User and stakeholder security can be enhanced through AI tools, which can take advantage of blockchain to open up new avenues for accessing and learning from data without taking ownership or control of that data. This can reduce risk for the organization and the stakeholders who provide the data. Both individual blockchain members and the organization or group in charge of setting governance rules and processes can benefit from building in privacy-related AI functionality (as early as possible) in the design of blockchain networks and processes.

Companies have implemented AI to create holistic views of customers by piecing together transactions from all customer touchpoints. Blockchain participants will have incentives to pull together integrated datasets by combining transactions for a single customer across all blockchain partners. This creates potential benefits for blockchain partners, but can also negatively affect the privacy of customers and other stakeholders for which this integration is possible.

In combination with options for identity protection through decentralization, AI can be used to combine personal data from blockchain participants and their stakeholders in a way that maintains information security and personal data privacy. Through these processes, user and stakeholder security can be enhanced and data sets and AI models can be improved.

We can identify four categories of stakeholders that can be affected by an organization's data transparency and privacy processes: (1) participants, whose data—both direct and indirect—are gathered; (2) victims, who are affected by decisions made using participant data; (3) users, who use participant data in their work; and (4) custodians, who manage and secure data. When AI can be used to manage access to data and to develop analytical models using that data, all stakeholders can benefit [15].

Table 1 summarizes a number of ways AI can be used in a blockchain setting to protect or increase privacy of user's personal data. This AI/Blockchain combination can increase system security by helping to detect attacks by bad actors, user security by sharing permissions and smart contracts, enable privacy-enhanced use of datasets through improved identity management and better data, and it can improve AI models through more varied, valid, and ethically-sourced data and better hypotheses. Each item in **Table 1** is described briefly below, followed by examples of use cases using this combination of technologies.

Computational intelligence (CI), a subset of AI can improve the Blockchain's attack resilience thus improving security of the system and ultimately the privacy of the data residing on the system. AI is rooted in hard computing techniques whereas computational intelligence is based on soft computing methods, which enable adaptation to a range of changing variables [16].

Computational intelligence, when combined with blockchain systems, can create more robust cryptographic functionality and ciphers thereby making it more difficult for cyber hackers to compromise systems even as computing power and efforts to hack these systems over time increases. Quite appropriately, [14] refer to the intersection of blockchain and AI as "blockchain intelligence". Additionally, AI algorithms can be built on blockchains to detect when a blockchain is under

System Security	• Malicious attack detection
	• More robust cryptographic functionality
User Security	• Users decide what data to share
	• Smart contracts can enforce established permissions
Datasets	• Improved identity masking and metadata
	• Cleaner and more accurate data
AI Models	• Broader scope and greater variety of data
	• Improved validity of data and models
	• Ethically sourced and permissioned
	• Careful construction of hypotheses

Table 1.
The role of AI in blockchain user privacy.

attack by continually monitoring blocks and activity on the chain. This technology increases trust in the system beyond what the native architecture provides [17, 18].

When blockchain participants have increased control over their own data, they have the potential to decide with which parties and for what purposes their data are shared. In order to collect participant data for use in an AI dataset, participant permissions will need to be obtained. This provides users with 'opt-in' control, rather than 'opt-out' and helps to ensure that personal data is used in ways that are consistent with the intentions of the owner. In some cases, which may become increasingly common if decentralized identity solutions are adopted, users can be compensated for providing their data to organizations seeking to utilize this data in traditional and AI decision models.

Smart contracts can also protect privacy. Permissions granted by users can be subject to complex rules embodied in smart contracts, which can enforce rules regarding the use of the data and can govern the granting and rescinding of participant data. AI can be used to scan contracts to identify participants who have or are likely to possess or provide data for desired uses.

The size and nature of datasets available in blockchain networks can also have implications for effective AI. Because many different organizations and stakeholders contribute to shared ledgers, the quantity and variability of data available for analysis can be much larger than for single-company databases. Larger datasets could enable more sophisticated identity-masking procedures, and metadata may be richer and more informative.

Because of the validation, security, timestamping and the append-only nature of blockchain ledgers, the data obtained are likely to be much cleaner and more accurate than when data are captured and maintained by many organizations in databases that are not immutable.

The ethical quality of data obtained will also be higher, and model developers and users can have increased confidence that they are following regulations. Because multi-dimensional user permissions can be granted and documented—and in some cases, enforced through smart contracts—organizations can use this data with less risk of privacy breaches. In addition, because user data can be collected using zero-knowledge proofs, complex analyses requiring specific user data can be performed and the necessary information captured and used without the need for accessing or possessing PII.

The use of blockchain data and artifacts can also result in higher quality analyses and outcomes. When data are clean and associated with clear metadata, the validity of the data is increased. Because each item in a data set is more trusted, error can be reduced and insights can be obtained through smaller data sets. When clean data are used to train AI models, those models will be more accurate, and the predictions and decisions made by those models will also be improved. Clean-training data can also be useful in validating non-blockchain data for use in AI models.

Finally, and perhaps most importantly, the hypotheses upon which AI models are built can improve, for several reasons. First, because participant permission must be obtained and possibly paid for, AI designers will need to develop clear designs that define the analyses to be performed and determine the type and amount of data needed for these analyses. This will require designers to be more aware of the universe of data that could inform these analyses and what is and is not available in distributed ledgers and personal-data files. This could help identify problems such as the lack of black faces in photo-categorizing algorithms before or during data collection.

PII may never be collected, and when it is, its use may be more intentional and usage agreements may be enforced by smart contracts. This enables more ethical approaches to gathering and managing data. AI models built using ethically-sourced and governed data can generate results that are actionable within pre-defined ethical and regulatory limits.

6.1 Emerging blockchain and AI industry uses cases

The 2019–2020 Covid 19 pandemic has prompted medical researchers and technologists to research ways to quickly gather intelligence around virus exposure and transmission as a way to combat the spread of the disease while maintaining personal privacy of users. Point-of-care diagnostics, which rely on rapid testing of patients that may have been exposed to the virus is proving to be an effective way of tracking the spread and reducing the impact of the disease. German based Pharmact AG has developed a rapid Covid 19 test that delivers results in roughly 20 minutes. This test can be used in point-of-care systems and combined with blockchain and AI to increase the speed of diagnosis and provide statistics on positive and negative results while maintaining security of personally-identifiable data. Data can be collected on blockchain infrastructure while taking advantage of the speed that AI affords to create an integrated platform that enables data from disparate sources to be analyzed. Information drawn from these systems can provide communities a powerful tool for combatting the spread of disease, reducing the burden of health care facilities and saving lives [19].

Many cities are working toward becoming "smart cities" by integrating AI and blockchain with other web 3.0 technologies such as internet of things (IoT) sensors and edge devices. Intelligent transportation systems are enabled by these technologies. Self-driving cars make use of IoT sensors to continuously monitor surrounding situations and even anticipate developments by using artificial intelligence. These cars can incorporate blockchain wallets that enable passengers to pay for rides, rentals, tolls, etc. without revealing personal data. By adding blockchain as an underlying architecture, cities and private companies can reduce the friction of renting or sharing autonomous vehicles by streamlining the process of procuring a ride. The peer to peer nature of blockchain reduces the number of people or businesses involved in the process, taking out expensive intermediaries, and reducing costs. These systems can also provide audit trails for both owners and renters, and enable rating and payment systems that maintain privacy for both parties. Data gathered by the vehicles can contribute to learning algorithms on the blockchain for increased security, scalability and efficiencies as well as improved transportation and sustainability for the city [20].

Smart home systems that preserve user privacy while contributing usage data for analysis can likewise benefit from the integration of blockchain and AI. Smart home systems are becoming popular and manufacturers increasingly enable connectivity between devices. These systems are valuable sources of consumer usage data. AI-enabled blockchain systems can be used to push machine learning and training processes to consumer's mobile devices and edge computing servers. Users can then submit locally-trained models for analysis, in some cases with option of adding noise that makes it very difficult to trace shared data back to individual consumers. Decentralized technologies enable analysis of locally generated data without this data being submitted to a centralized server [21].

These use cases exemplify some of the ways blockchain and AI are being used to accomplish objectives while maintaining personal data privacy. New use cases continue to be developed as technologists and user communities recognize the possibilities for systems that provide both functionality and privacy.

7. Conclusion

Blockchain and AI technologies are improving at a rapid pace and enabling possibilities for sharing and combining data in ways not previously envisioned.

At the same time, advances in these technologies provide new possibilities for the ethical use of data. Personal data, when shared, present a conundrum for firms and individuals, which can provide valuable benefits but can also create great risks and costs for both the individual and the organizations with which individual data are shared. Blockchain provides new mechanisms, such as decentralized identities and zero-knowledge proofs, that enable data to be shared in ways that maintain the privacy of the individual and allow users to maintain control over their own data. These advances can provide both increased cybersecurity and more ethical use of personal data. Blockchain participants can realize these outcomes through careful development of governance frameworks and mechanisms.

Publication of this chapter in an open access book was funded by the Portland State University Library's Open Access Fund.

Author details

Stanton Heister* and Kristi Yuthas
Portland State University, Portland, OR, USA

*Address all correspondence to: stanton.heister@pdx.edu

IntechOpen

References

[1] RiskBased Security. 2020 Year End Report: Data Breach Quickview. [Internet]. 2021. Available from: https://pages.riskbasedsecurity. com/hubfs/Reports/2020/2020%20 Year%20End%20Data%20Breach%20 QuickView%20Report.pdf [Accessed: 2021-01-12]

[2] Kellerman R. Five of the Biggest Data Breaches of the 21st Century. *STAGE2DATA*. [Internet]. 2020. Available from: https://www.stage2data. com/five-of-the-biggest-data-breaches- of-the-21st-century/ [Accessed: 2020-12-17]

[3] Fruhlinger J. Equifax Data Breach FAQ: What Happened, Who was Affected, What was the Impact? [Internet]. 2020. Available from: https:// www.csoonline.com/article/3444488/ equifax-data-breach-faq-what- happened-who-was-affected-what-was- the-impact.html [Accessed: 2020-11-21]

[4] Statista, Figure 1. Cybersecurity Breaches and Record Exposure [Internet]. 2020. Available from: https:// www.statista.com/statistics/273550/ data-breaches-recorded-in-the- united-states-by-number-of-breaches- and-records-exposed/ [Accessed: 2020-11-12]

[5] Heister S, Yuthas K. Technology in Society: The Blockchain and How it Can Influence Conceptions of the Self. 2020. (60) https://doi.org/10.1016/j. techsoc.2019.101218

[6] Grimes R. What is Personally Identifiable Information (PII)? How to Protect it Under GDPR. [Internet]. 2019. Available from: https://www.csoonline. com/article/3215864/how-to-protect- personally-identifiable-information-pii- under-gdpr.html [Accessed: 2020-11-21]

[7] Uribe D, Waters G. Privacy Laws: Genomic Data and Non-Fungible Tokens. *The Journal of the British Blockchain Association*. 2020. (3) https:// doi.org/10.31585/jbba-3-2-(5)2020

[8] Yoon J. Democratic Senators Introduce the Consumer Online Privacy Rights Act. [Internet]. 2019. Available from: https://www.insideprivacy.com/ united-states/congress/democratic- senators-introduce-the-consumer- online-privacy-rights-act/ [Accessed: 2021-01-08]

[9] Consensys. Busting the Myth of Private Blockchains. [Internet]. 2020. Available from: https://consensys.net/ enterprise-ethereum/best-blockchain- for-business/busting-the-myth- of-private-blockchains/ [Accessed: 2020-12-12]

[10] Microsoft. Decentralized Identity. [Internet]. Available from: https://www. microsoft.com/en-us/security/business/ identity/own-your-identity [Accessed: 2020-12-12]

[11] ElGayyar M, ElYamany H, Grolinger K, Capretz M, Mir S. Blockchain-Based Federated Identity and Auditing. *International Journal of Blockchains and Cryptocurrencies*. 2020.p. 179-205

[12] Lesavre L, Varin P, Mell P, Davidson M, Shook J. A Taxonomic Approach to Understanding Emerging Blockchain Identity Management Systems. [Internet]. 2019. *COMPUTER SECURITY RESOURCE CENTER*. Available from: https://csrc.nist.gov/publications/ detail/white-paper/2019/07/09/a- taxonomic-approach-to-understanding- emerging-blockchain-idms/draft [Accessed: 2020-11-17]

[13] Wikipedia. Zero-Knowledge Proof. *Wikipedia*. [Internet]. 2020. Available from: https://en.wikipedia.org/wiki/

Zero-knowledge_proof [Accessed: 2021-01-18]

[14] Zheng Z, Dai H, Wu J. Blockchain Intelligence: When Blockchain Meets Artificial Intelligence 2020. *arXiv preprint arXiv:1912.06485.*

[15] Bertino E, Kundu A, Sura Z. Data Transparency with Blockchain and AI Ethics. Data and Information Quality; 2019. https://doi.org/10.1145/3312750

[16] Wikipedia. Computational intelligence. *Wikipedia.* [Internet]. Available from: https://en.wikipedia. org/wiki/Computational_ intelligence#:~:text=According%20 to%20Bezdek%20(1994)%2C,a%20 subset%20of%20Artificial%20 Intelligence.&text=Crisp%20logic%20 is%20a%20part,be%20partially%20 in%20a%20set. [Accessed: 2021-01-18]

[17] Marwala T, Xing B. Blockchain and Artificial Intelligence 2018. Arvix preprint arXiv:1802.04451.

[18] Salah K, Rehman M, Nizamuddin N, Al-Fuqaha A. Blockchain for AI: Review and Open Research IEEE; 2019. (7) p.10127-10149. DOI: 10.1109/ ACCESS.2018.2890507

[19] Mashamba-Thompson T, Crayton E. Self-Testing: Blockchain and Artificial Intelligence Technology for Novel Coronavirus Disease 2019. Diagnostics. 2020. (10) 198. https://doi.org/10.3390/ diagnostics10040198

[20] Singh S, Sharma P, Yoon B, Shojafar M, Cho G, Ra I. Convergence of Blockchain and Artificial Intelligence in IoT Network for the Sustainable Smart City. 2020. *Sustainable Cities and Society.* (63) art. no. 102364

[21] Zhao Y, Zhao J, Jiang L, Tan R, Niyato D, Li A, Lyu L, Liu Y. Privacy-Preserving Blockchain-Based Federated Learning for IoT Devices. IEEE Internet of Things Journal. 2020.

Chapter 4

Blockchain-Empowered Mobile Edge Intelligence, Machine Learning and Secure Data Sharing

Yao Du, Shuxiao Miao, Zitian Tong, Victoria Lemieux and Zehua Wang

Abstract

Driven by recent advancements in machine learning, mobile edge computing (MEC) and the Internet of things (IoT), artificial intelligence (AI) has become an emerging technology. Traditional machine learning approaches require the training data to be collected and processed in centralized servers. With the advent of new decentralized machine learning approaches and mobile edge computing, the IoT on-device data training has now become possible. To realize AI at the edge of the network, IoT devices can offload training tasks to MEC servers. However, those distributed frameworks of edge intelligence also introduce some new challenges, such as user privacy and data security. To handle these problems, blockchain has been considered as a promising solution. As a distributed smart ledger, blockchain is renowned for high scalability, privacy-preserving, and decentralization. This technology is also featured with automated script execution and immutable data records in a trusted manner. In recent years, as quantum computers become more and more promising, blockchain is also facing potential threats from quantum algorithms. In this chapter, we provide an overview of the current state-of-the-art in these cutting-edge technologies by summarizing the available literature in the research field of blockchain-based MEC, machine learning, secure data sharing, and basic introduction of post-quantum blockchain. We also discuss the real-world use cases and outline the challenges of blockchain-empowered intelligence.

Keywords: blockchain technology, mobile edge computing (MEC), distributed machine learning, internet of things (IoT), data security and privacy

1. Introduction

In the past few years, machine learning and blockchain have been known as two of the most emerging research areas. Machine learning is the practice of building learning models on the computers to parse data, and provide human-like predictions or decisions for some real-world problems. Blockchain, on the other hand, has the capability to store and process data, preserve data integrity, and govern peers accessibility without needing any centralized administration. Those two research areas are heavily data driven and each of those technologies has its own advantages and bottlenecks. In this chapter, we review some novel research on combining blockchain and machine learning, and identify how their short-comings can be addressed by merging these different ecosystems. Several machine learning techniques such as supervised machine learning, deep reinforcement learning and federated learning are considered as good alternative solutions to the Blockchain related research. We also discuss how the researchers make these two technologies work collaboratively to solve some real-world problems.

The Internet of things (IoT) is a well-known technology for research and industry. Devices in this network can still sense and respond to the environment without users' intervention. Enabling artificial intelligence (AI) in IoT has emerged as a hot research topic [1]. However, machine learning is a kind of computational task which is a heavy workload for the IoT devices (IoTDs). Usually, these low-cost IoT devices are battery-powered devices. On the one hand, computational tasks execution (e.g., training machine learning models) consumes considerable energy. On the other hand, the required powerful microchips are not suitable for IoTDs with compact physical size.

Mobile edge computing (MEC) is a solution to the above challenges. By offloading complex learning tasks to the edge of Internet, IoTDs could perform machine learning algorithms to realize AI. The original MEC was proposed by ETSI in 2014. The description of MEC was "A new platform provides IT and cloud computing capabilities within the Radio Access Network (RAN) in close proximity to mobile subscribers" [2]. We focus on the edge computing within the RAN in this chapter because it has been a standard across different industries. It will help readers to learn blockchain-enabled mobile edge intelligence in the most practical scenario. However, security and privacy issues must still be considered [3]. The IoT data leakage may lead to malicious attacks on individuals. Fortunately, blockchain has potential and is suitable for MEC [4]. The integration of MEC and blockchain is a win-win solution. For one thing, blockchain provides MEC with data security and privacy. For another, MEC can improve blockchain's scalability and effectiveness.

The main contributions of this chapter are listed as follows:

- We first focus on the security and privacy-preserving features of blockchain. The consensus mechanisms, permissioned blockchain and zero-knowledge proof are jointly introduced to give a general understanding for readers;

- We describe the blockchain-enabled mobile edge intelligence in the scenario of IoT systems. The potential of blockchain in edge intelligence is included;

- The combination of blockchain and AI is further given in a two-way manner. Real world applications of blockchain in AI are summarized;

- Discussions of blockchain's threat are illustrated for future research. For example, the threat of quantum computing and its related research is surveyed in this chapter.

The rest of this chapter is organized as follows. In Section 2, we introduce the security and privacy-preserving features of blockchain. Next, the blockchain-enabled edge intelligence is discussed in Section 3. Then, some blockchain and machine learning combined research is summarized in Section 4. Finally, we discuss some threats for blockchain-enabled systems and conclude this chapter.

2. The security and privacy-preserving features of Blockchain

Blockchain is famous for its security, and the most well-known social experiment of blockchain today is Bitcoin. As a revolutionary invention, Bitcoin certainly caught many investor's attention. One index that can be used to determine the popularity of Bitcoin is the total number of wallets created on the Bitcoin networks[1]. This index is currently at record high (around 65.015 million). Unlike traditional banking systems where a person can only have a limited number of bank accounts, Bitcoin allows users to create accounts with just a few commands/clicks without the involvement of any government issued identity verification process. Therefore, we would not be able to find out how many people tried Bitcoins or engaged in the Bitcoin network service. There is another number that can show the impact of Bitcoin, which is the total hashrate (TH/s) of the Bitcoin network[2], this number is also at its record high, about 153.019 million (TH/s) now. Mining hashrate is one of the key security metrics of the Bitcoin network. The bigger the hashrate number or hash power in a network, the greater its security and its overall resistance to attacks. There is no way for us to calculate the actual hashrate on the network. However, from the block difficulty, we can give an estimate of its total hash power [5]. Hash power is delivered by Bitcoin miners, whose computers join the blockchain network to compute the problems together. This mechanism is often called proof-of-work (PoW) and it costs a lot of power and electricity. A study published on Nature Climate Change in 2018 estimated that Bitcoin mining alone could push up global warming by 2 degrees Celsius [6].

At this moment, a single Bitcoin is worth around 40,000 US dollars. The questions are, what are people buying it for? Are there any reasons why mining mechanisms are so energy hungry? Are there any alternative ways to design systems? In the following paragraphs, we will explain to you how Bitcoin achieves its state-of-the-art secure distributed ledger and how it allows people on the boundary of trust to work together. Next, we will introduce you to another concept and discuss how 5G edge devices could benefit from blockchain security.

2.1 Security feature

2.1.1 Proof-of-Work (PoW Consensus Algorithm)

PoW is the underlying consensus algorithm of Bitcoin [7]. A consensus algorithm in computer science is a process used to achieve agreement on a single data value among distributed processes or systems [8]. This term is commonly used by distributed systems. It explains that in the modern computing era, how multiple servers could work together with high levels of security and fewer errors. In more detail, some servers (or nodes) may fail or may be unreliable in other ways (e.g., being hacked, losing data, running in idle). Therefore, consensus algorithms must be fault tolerant and resilient. Consensus mechanisms function a bit like constitutions in the human world - guiding decisions about what's acceptable among interacting parties. This is the core of a blockchain system. PoW is commonly recognized as a secure consensus algorithm [7]. It's a consensus mechanism that heavily relies on computing power and cryptographic hash function (also known as CHF, a mathematical algorithm that maps data of arbitrary size to a bit array of a fixed size). Before diving into this consensus model, one needs to be familiar with the following two concepts, the Crash-Fault Tolerance (CFT) and the Byzantine-Fault Tolerance (BFT).

The Crash-Fault Tolerance (CFT). Just as it states in its name, it can be resilient toward crash/halt events. Suppose that your system has been damaged or lost

---

[1] https://www.blockchain.com/

[2] One trillion (1,000,000,000,000) hashes per second.

connection, the CFT based system will still function and give the result that you expect [9]. The CFT fault model has been discussed by academics for a number of years long time and is mature in industrial use cases. Most cloud based companies implemented different CFT methods to prevent the critical problems. We still occasionally hear news about collapse/maintenance of servers of Amazon, Alibaba or Microsoft and developers make some open source or free projects to show you the real-time status of their servers (e.g., the downdetector[3]).

Byzantine-Fault Tolerance (BFT) is more complicated than CFT. The name Byzantine-Fault Tolerance is derived from a paper published by Leslie Lamport, Robert Shostak and Marshall Pease on SRI International called The Byzantine Generals Problems [10].

Converting the story above into a computer system use case, a distributed system should be resilient to the case when a small portion of the computers in the network are malicious. Every non-malicious entity has the same status (including action). Companies' server systems are mostly CFT because they set up each node/server in the network and they are very confident that the chances of having malicious nodes are low. Furthermore, they have backup systems that can recover from the previous loss. However for Bitcoin, everyone can hop-on and hop-off the Bitcoin chain. So how do we protect the network and the ledger system's integrity, making sure no one is taking extra money that does not belong to them or preventing one's money from being stolen [7]. This is a typical BFT question.

The PoW solves the problem of determining representation in majority decision making. It's a one-CPU-one-vote proposed by Satoshi Nakamoto [7], the pseudonymous founder of Bitcoin. Around every 10 minutes, the network will wrap around all transactions that happened within the 10 minutes. As illustrated in **Figure 1**, it uses a cryptographic hash function to hash previous block's hash, Nonce and transactions together to form a new hash block. Each node will try different nonces (numbers) to find a certain number of zeros in front of the hash result (mining difficulty). This process is called mining, it is power consuming and no less energy consuming, mathematically proven secure short cut has been found yet.

The beauty of PoW is that the result can be instantly verified but it will be very difficult to tamper with unless the malicious nodes occupy 51 percent of the total computing power of the network [11]. With increasing numbers of miners joining the mining and significant numbers of investments in this area, the difficulty of tampering with the Bitcoin network becomes harder and harder. However, a lot of criticisms about blockchain also arose in the past decades, especially about its efficiency, energy consumption and its economic model. Environmentalists disagree with this mechanism due to the fact that the annual electricity consumption of the Bitcoin network is nearly 120 gigawatts (GW) per second. Equivalent to 49.440 wind turbines (412 turbines per GW) when generating power at peak production per second [12]. There is another focus on Bitcoin PoW mechanisms which is about the ASIC miners. Since the Bitcoin mining mechanism is like solving a math puzzle,

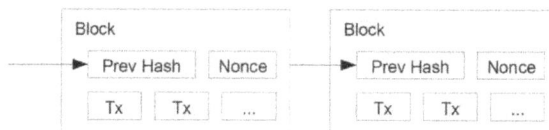

Figure 1.
Blockchain Structure [7].

---

[3] https://www.isitdownrightnow.com/

developers moved from using CPU to using Application-Specific Integrated Circuit (ASIC) machines to mine Bitcoin. This shift significantly increased the Bitcoin mining difficulty since ASIC is a dedicated computing circuit that is purely designed for mining purposes [13]. Therefore, the ASIC's performance is much better than the General Purposes Computing Unit (CPU). This brings a huge barrier for beginners to join the mining, and it will further enhance the Bitcoin mining centralization, not to mention that the Bitcoin miner market is in a relatively monopoly situation, as 66% of market share of miners are occupied by Bitmain[4].

2.1.2 Alternative consensus mechanisms

Due to the above drawbacks and concerns over PoW mechanisms, designers have developed a new consensus mechanism called Proof-of-Stake (PoS) where miners have been incentivized based on how much "work (hash power)" they have contributed. In contrast, the PoS mechanism asks miners to bet tokens in order to participate in the new block generation. This concept was first discussed in 2012 by Peercoin [14, 15], and had later become a popular discussion and experiment among many cryptocurrencies.

Compared with the PoW mechanism, PoS is simpler. The more you stake, the higher the influence you have on determining the next block [14]. The assumption is based on the premise that high staking nodes are less likely to lie because of high losses for them if they tried to tamper with the chain. Under this scenario, nodes are no longer needed to solve cryptographic puzzles. Apparently, compared with PoW, PoS is more environmentally friendly. Besides, less computation will further enhance the network transaction speed as well as the throughput [16]. There are currently two types of PoS consensus models: Chain-based Proof-of-Stake and Consortium Consensus model [14]. Chain-based Proof-of-Stake chooses availability over consistency [17]. The algorithm pseudo-randomly selects a validator during each time slot. Validators will then have the right to create a single block and point it back to the previous longest chain. Consortium Consensus Model chooses consistency over availability. During each voting procedure, every node counts proportionally to the stake it bets.

Apparently, the richer nodes have higher chances of getting the reward. In order to avoid the monopoly situation, a random selection model is required for most PoS algorithms. Each token has their own token economy and some of them include token age design, where tokens being staked for a longer time will be more likely to be selected. In general, the network wants holders to stake validation tokens as long as possible and act as hosts.

With rapid adoption of 5G networks by different countries and regions, edge computing and edge devices will play an important role in the network. Blockchain is currently the state-of-the-art secure network system and has many use cases in edge computing [18]. In the upcoming 5G era, more and more devices will have access to the network, since one of the main goals of 5G is to support the IoT [19, 20]. As predicted, data transmission rate and volume will exponentially grow and thus to have a secure communication channel and a consensus algorithm for edge devices have been identified in recent research papers. Edge devices for 5G network are often not designed for doing heavy computation. Thus, a new consensus model is needed for the 5G era. PoS as one of the successful consensus models has potential in the IoT era due to its simpler structure, high security feature and completely decentralized characteristics. Besides PoS, Delegated Proof-of-Stake

---

[4] https://www.bitmain.com/

(DPoS) is another possible consensus algorithm for 5G use cases with greater scalability and faster transactions. However, DPoS is not completely decentralized; it is an intermediate solution finding the balance between centralization and decentralization [21].

2.2 Privacy-preserving features

In 2020, a Netfliex documentary called The Social Dilemma raised awareness about risks to our personal privacy. The Social Dilemma exposes audiences to shocking facts about how social media apps are currently using intelligence algorithms to control user's behaviours, as a result, making users addicted to their own content, and gathering user's data to target users with ads without any regulatory supervision [22].

Traditional banking systems often require many documents to set up one account. However, registering a public blockchain network normally does not require any identity verification and there is no limitation on how many accounts that you can build and there is no cost associated with making an account. Bitcoin (BTC) and Ethereum (ETH) are currently the most popular public blockchain networks[5]. When you send Ether (the cryptocurrency in Ethereum) on Ethereum platform, the sender and receiver are both just wallet addresses, hashes of a public key. Each transaction will be broadcasted on the mainnet.

2.2.1 Permissioned Blockchain network

Blockchain has many user scenarios. As for Bitcoin, it brings people to work together even if they are all on the boundary of trust. Users in the Bitcoin network have no need to trust each other at the beginning and anyone can hop-on and hop-off the network. Public blockchain networks are mostly targeting monetary systems and have their own token economics. However, public blockchain networks are often very slow and cannot be used in specific scenarios due to complex consensus algorithms to prevent malicious attacks. To make blockchain networks be more useful and specific to each use case, developers proposed permissioned blockchain network ideas. Permissioned blockchains usually involve a consortium of organizations who are in charge of verifying the transition history instead of asking pseudonymous miners to participate in the mining process [23]. Permissioned blockchains are often popular in the industries that rely on digital data, for example, supply chain management, liquidation in the financial industry, manufacturing industry. These industries take data security, data privacy and role definition seriously and are keen on pursuing higher efficiency. The operation of these industries is often similar to a chain reaction; one mistake in one process will cause sequential reaction in the following processes.

Public blockchain networks are open to anyone, where permissioned blockchains require identity verification as an extra security layer. In a nutshell, nodes on permissioned blockchains are verified and their roles in the network are predefined.

Permissioned networks can be used to protect sensitive data. Public blockchain networks set all users/nodes with equal amounts of power. In contrast, a permissioned blockchain network can have a more complicated internal structure to ensure data security, and access controls to specify that only nodes with permission can retrieve.

---

[5] https://coinmarketcap.com/, a website to track the market cap of all cryptocurrencies

2.2.2 Zero-knowledge proof and zk-SNARK

People may wonder whether there are mechanisms that can be used by a public blockchain network to hide some transactions or protect one's address from being traced. Zero-knowledge proof (ZKP) is a proof that allows a prover to prove the knowledge of a secret to a verifier without revealing it. The verifier should receive no knowledge before verification and after the verification [24]. This sounds unreal because in the real world, we gain trust in third parties through revealing our private messages or information such as date of birth, secret key or password. There is a famous story on ZKP. It was about two mathematicians who both claimed that they found the solution equation of a formula. However, Neither of them wanted to reveal it to the public. They later conducted a competition in which both drafted a question for their competitor and they would solve the question from each other by using their own method and to show their solutions. Verifiers only needed to check whether their solutions were correct to determine who actually found the solution equation of that formula. As a result, neither party revealed any information to the verifier, but the verifier still had sufficient evidence to determine who actually knew the solution.

Generally speaking, ZKP has two categories, the interactive and non-interactive ZKPs. Interactive ZKP is more intuitive. It requires intervention between individuals (or computer systems) to prove their knowledge and the individual validating the proof. For example, the method used by the mathematician competition is based on interactive ZKP. Interactive ZKP already has many applications in the communication industry, and such a system requires a stable and continuous communication channel. Non-interactive proof requires none of that, it takes less time and only one message is enough [25]. It's more efficient and can be optimized for IoT systems. Zero-Knowledge Succinct Non-Interactive Argument of Knowledge (zk-SNARK) is a type of non-interactive ZKP and has been used by cryptocurrency Zcash as its core privacy-preserving mechanism. In a nutshell, Zcash can hide some transactions to make the blockchain more privacy focused, and external parties will not be able to trace many accounts' transaction history. However, researchers from Carnegie Mellon University found that 99.9% of Zcash and 30% of Monero (another privacy-preserving token) transactions were traceable because users may not use them properly [26, 27].

3. Blockchain-enabled edge intelligence for IoT

In this section, we introduce the potential and applications of blockchain in edge intelligence. To be specific, we aim at describing how AI could be implemented on the edge of the Internet and how blockchain could improve the mobile edge intelligence. We first introduce offloading strategies and the MEC architecture for readers. Then, we list how blockchain can improve mobile edge intelligence in terms of data security. Finally, we describe the blockchain-enabled resources allocation and market trading in the mobile edge intelligence systems with the constrains of energy supply, computational power, and the size of training data.

3.1 Tasks offloading in Blockchain systems

Although the PoW can secure transaction records in Bitcoin and similar blockchain networks, it is still very challenging to implement this kind of consensus mechanism for securing AI applications because it is a computational intensive task. To be specific, realizing blockchain-enabled AI for edge devices (i.e., edge intelligence) requires two kinds of tasks to be executed. On one hand, mining process is necessary for establishing consensus among distributed IoTDs. On the other hand, data training will be performed at each IoTD in a decentralized manner.

However, IoTDs are energy-constrained and unable to process complex computational tasks. Fortunately, edge computing is capable of handling this issue. The main idea of realizing blockchain-enabled AI on the edge network is to use different offloading strategies for IoTDs. As illustrated in **Figure 2**, one can see that computing tasks can be offloaded from IoTDs to MEC servers located in different places, including small-cell base stations, cellular base stations, and the cloud data center [28]. As the distance between IoTD and MEC server becomes shorter, the computing capability decreases.

However, data leakage and other security issues impose great challenges to MEC, especially for MEC-enabled blockchain and AI applications. In [3], authors discussed the privacy issues in MEC-enabled blockchain networks. Data processing and mining tasks were offloaded to nearby servers. Moreover, the privacy level was modeled in this paper. The trade-off and optimization among energy consumption, privacy, and latency were jointly considered. Furthermore, [29] investigated the trust mechanism for edge network by using blockchain technology. Selfish edge attacks were discussed in this paper. The selfish attack means MEC servers provided the IoTDs with fake service and less computational resources. To deal with this attack, the authors explored the blockchain-based reputation record system, wherein selection of the miner relies upon the reputation of the MEC servers.

As the growth of IoTDs and AI applications becomes explosive, it becomes challenging to coordinate tasks offloading in wireless networks, especially for the ultra-dense wireless network [30]. To deal with this problem, authors in [30] proposed a decentralized platform based on blockchain. Computational tasks were first published and recorded in this platform, then user matching was evaluated and conducted based on the tradeoff between service latency and energy consumption. Moreover, offloading mode selection was discussed in [31], including offload tasks to a nearby server or a group of users. The content caching strategy was studied to handle the traffic issue in blockchain wireless networks. Furthermore, content caching could be used to extend the block capability of a blockchain platform. To be specific, hashed blocks were cached in MEC servers [32]. Then, images and even videos could be stored into the block for AI applications. Authors in [33] further optimized and proposed a block size adaptation scheme for video transcoding.

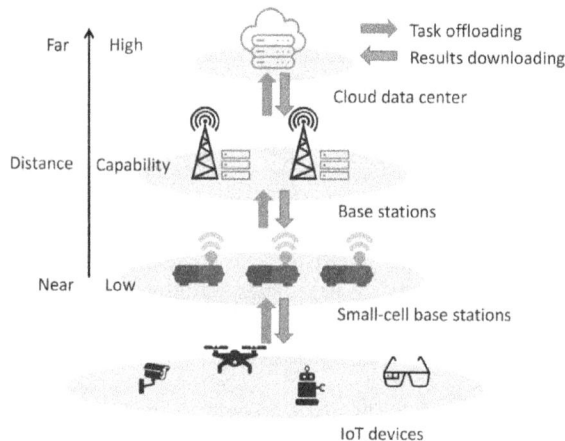

Figure 2.
MEC architecture for IoT devices.

Another issue is cooperation incentive. As mentioned above, nearby MEC servers have limited resources to share. The cooperation computation offloading research was discussed in [34]. To incentivize and establish this kind cooperation, a coin loaning strategy was given for IoTDs in [35]. Besides, most related research ignored the real need of IoTDs and blockchain-enabled AI applications. For example, fast transaction writing and uploading are critical for low-latency applications [36]. In another scenario, the revenue (e.g., tokens) may be treated as the first priority in the computing cooperation. Moreover, the scalability and efficiency should be considered in offloading strategies. To solve this issue, blockchain technology and the directed acyclic graph were explored in [37].

3.2 Blockchain-enabled data security for mobile edge intelligence

IoT data contains sensitive information related to individuals. Therefore, IoT data security is critical in establishing AI applications based on IoTDs. As described in the previous part, blockchain-based MEC is a key solution to enable AI for IoTDs. To secure IoT data and MEC, blockchain is a promising strategy. To be specific, authentication, secure communication, data privacy, and data integrity are four main strategies to enable IoT securtiy in the scenario of MEC.

Although blockchain-enabled edge intelligence is a cutting-edge technology to enable AI applications for IoT systems, it is vulnerable when facing malicious attacks. Authentication of identities in blockchain-enabled MEC system was discussed in [38]. The authors proposed a digital validation strategy based on group signature scheme. In this way, the identities of block creators were verified and authenticated to prevent edge intelligence from false records. Moreover, to manage privacy and data leakage issues, authors in [39] discussed the authentication for federated learning (FL) peers in the edge computing context. By exploring blockchain technology and smart contract, the FL and differential privacy technique were proposed in this article.

Secure data sharing is another aspect of data security for edge intelligence. The basic idea of secure data sharing is that data should be stored and shared in a trusted way. Consortium blockchain was discussed in [40] for the secure and efficient data sharing. Different from a public blockchain network, the consensus process is performed on a group of pre-determined edge nodes in consortium blockchain. The proposed model contained two kinds of smart contracts, including data storage smart contract and information sharing smart contract, allowing the auditing and governance of data sharing. Besides, the data sharing problem was transformed into a machine learning problem in [41]. FL was explored in this article to share the learning model rather than raw data.

The integrity of data should be considered in edge intelligence. As illustrated in **Figure 3**, any false information may do harm to the global ML model. For example, poisoning attack and input attack are two major types of issues in data integrity in AI application. The first one normally occurs in the initial stage of AI model training. By manipulating training data, the AI model would be ineffective. The latter one tends to use manipulated data to shape and affect the AI model output in a way desired by attackers. The design of blockchains naturally protects data integrity because any tampering of previous data recorded are not permissioned. MEC servers were used to validate and store blockchain data in [42] for data integrity. The data acquisition process for IoT system was discussed in this article. To be specific, the identity of a data sender was verified in this process. IoT data were recorded and stored in blockchain only if the validation was successful. Additionally, verifiable integrity of IoT data could be realized by blockchain [43].

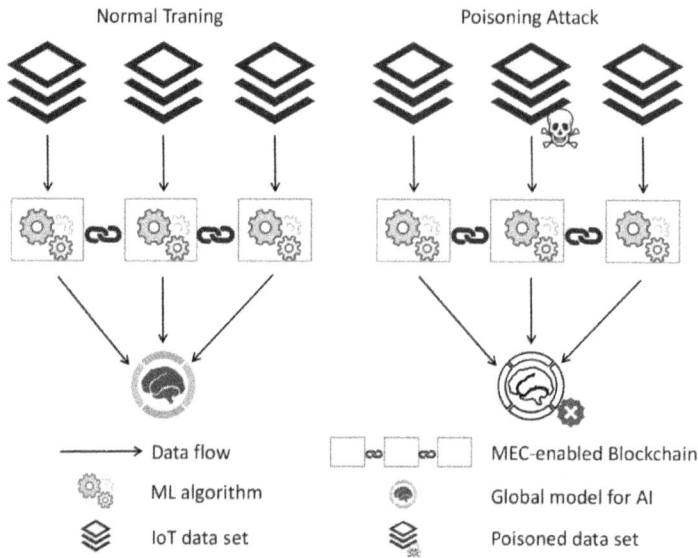

Figure 3.
Poisoning attack on blockchain-enable edge intelligence.

As we have discussed in the previous part, protection of privacy presents great challenges for mobile edge computing. Sensitive data should not be shared to any trustless third parties. The decentralization of privacy was discussed in [44]. The off-chain storage technique was used in the blockchain for privacy-preserving purposes. Furthermore, the topology of the edge intelligence network is another kind of sensitive information. A heterogeneous MEC system was proposed in [45] to provide MEC network topology with protections. FL was used in [41] to solve the privacy issue in data sharing. Differential strategy was further integrated into FL to prevent the leakage of sensitive IoT information. In terms of industrial IoT systems, authors in [46] explored the privacy issue in industry 4.0. ML models were trained on the sensitive data in industrial IoT systems. Therefore, the establishment of trustworthiness and privacy-preservation was a critical aspect in designing AI applications for industry 4.0 aspect. By jointly exploring the Ethereum blockchain, smart contracts, differential privacy, and FL, the authors proposed a novel blockchain named PriModChain to handle the privacy and security issues in industrial IoT systems.

3.3 Blockchain-based market for mobile edge intelligence

In the social layer, blockchain is a promising technology for peer-to-peer resources trading. As is shown in **Figure 4**, energy, information, and computing power are three major resources in blockchain-enabled markets. As we have discussed, MEC servers are more powerful than the IoTDs in terms of storage and computing power. However, IoTDs are far more than MEC servers. That means resources are still very limited on the edge of Internet. Therefore, resources allocation is critical for blockchain-enabled edge intelligence. Fortunately, blockchain can establish an open market among IoTDs and MEC serves, enabling resources trading according to the need in system level.

Energy trading is important and useful in IoT systems. This is because IoTDs are energy-constrained. Different from the MEC servers with the constant power supply, IoTDs are battery-powered and not very convenient to recharge.

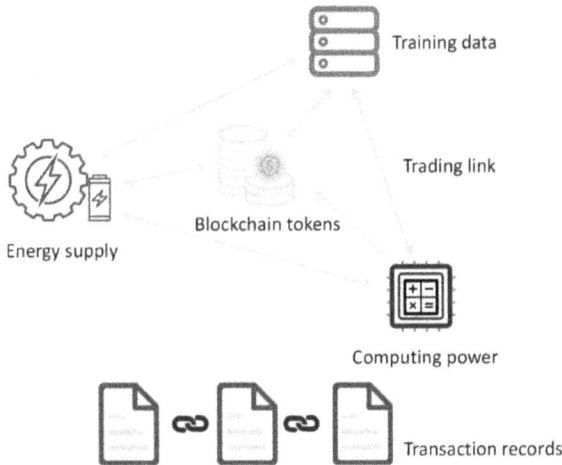

Figure 4.
Blockchain-enabled resources market for edge intelligence.

Energy-knowledge trading was discussed in [47]. On-device AI applications would be useless when IoTDs' battery is exhausted. Therefore, authors in this paper proposed a wireless way to power IoTDs. A permissioned blockchain was used for peer-to-peer (P2P) resources trading between energy power and training data sets.

Marketing is not only a platform to trade resources but also an effective solution to incentivize IoTDs and MEC servers into this blockchain-enabled platform. FL training market was studied in [48]. The idea of using blockchain is to establish a decentralized and fair market among MEC servers groups. A smart contract based resources trading market was further proposed in this article to ensure automatic transactions.

Content market is another topic for blockchain-enabled edge intelligence. To be specific, video transcoding and content delivery were investigated in [49]. A decentralized market was established by blockchain among trustless entities in a content delivery network. Content price, offloading cost, and content quality were jointly considered in this article. Furthermore, the willingness of MEC caching was discussed in [50]. In general, blockchain incited MEC servers by satisfying their expected rewards.

Last but not least, data is another type of digital asset, especially for AI applications. To be specific, data ownership is critical in the performance evaluation of AI training. Authors in [51] investigated the data ownership in AI training. Transactions among data owners, AI developers, and service providers were recorded in the proposed blockchain system. In this way, trading and data usage actions became traceable and verifiable. Thus, the ownership of training data sets was well-preserved. Nevertheless, for some types of sensitive transactions, where the existence of a transnational relationship between two parties may need to remain private, residual privacy concerns remain when all intervention is recorded on the ledger.

4. Blockchain and machine learning combined research

4.1 Machine learning for Blockchain industry

4.1.1 Blockchain security attack detection

One thing the public are concerned with respect to blockchain is its security performance. Although blockchain utilizes cryptography and consensus to enforce network security and privacy, it is still not immune to potential attacks. In 2017, the Bitcoin researchers [5] found out that the Bitcoin network is vulnerable to some state of the art attacks, even though it has been successfully running for 8 years. In 2019, some vulnerabilities in the Ethereum network were exposed and it was reported that the network has experienced several attacks such as 51% attacks and data breaching attacks [52].

Machine learning has been considered as one of the tools to improve the blockchain security. Scicchitano *et al.* in [53] introduced an unsupervised machine learning approach to identify anomalies in the activities of the blockchain network. The proposed anomaly detection system constructs a neural encoder-decoder model and the model is capable of summarizing the status of the ledger sequence-by-sequence. The system has the ability to detect the difference of the statuses between standard situation and anomalous situation and trigger the alert accordingly.

Somdip Dey [54] was interested in improving the blockchain consensus mechanism. By utilizing game theory and supervised machine learning algorithm, an improved Proof-of-Work consensus is introduced to prevent any quantifiable attacks. By analyzing the attacker's activities and rewards, a utility/payoff function can be derived and fed into a supervised machine learning model. This machine learning model has the ability to detect whether an attack is bound to happen or not - based on the value of the commodity/service. If the attack is likely to happen, the machine learning agent has the ability to prevent the blockchain confirmation until a new block of fair transactions is generated again.

Hou *et al.* [55] proposed a framework called SquirRL. This is a deep reinforcement learning framework that can be used to analyze the blockchain rewards. Even though SquirRL is used to detect the adversary activities in the network, the framework can automate vulnerability detection in the blockchain incentive mechanisms. When the theoretical analysis is infeasible, SquirRL can serve as a powerful tool for the blockchain engineers to verify the protocol designs during their development phrases.

4.1.2 Cryptocurrency and mining

Thanks to the blockchain and cryptography, the emergence of cryptocurrency has drawn a lot of attention. Unlike fiat money and stocks, the cryptocurrencies have shown significantly unstable fluctuations and have disrupted the investment industries. Researchers have made steady progress on how to improve the profit in the cryptocurrency by applying machine learning models to analyze market performance or network data.

One main direction in this research area is to utilize the machine learning models to predict the prices of cryptocurrencies. However, the data source and the detail techniques may vary. Kim *et al.* [56] introduced a method that can help predicting the cryptocurrencies fluctuations. The proposed method collects user online posts and comments that are related to the cryptocurrency market activities, and conducts an association analysis between the collected data and the fluctuations in the prices of the cryptocurrencies. The final model shows about 74% weighted average precision in the Bitcoin and Ethereum markets. Madan *et al.* [57] intended to automate Bitcoin trading via supervised machine learning algorithms by using random forest and binomial logistic regression to support vector machine. Their learning method is trained with the Bitcoin price index and the final result achieves above 55% precision. Jang and Lee [58] used Bayesian Neural Network algorithm to train the supervised learning model. The training data for their empirical study includes cryptocurrency market prices and volumes, blockchain attributes, financial stock market information and global currency ratio. Their research presents a promising result of anticipating the Bitcoin price time series and explaining the high volatility of the Bitcoin market. McNally *et al.* [59] assembled two different deep learning models to forecast Bitcoin price - with Recurrent Neural Network (RNN) and Long-Short Term Memory (LSTM) algorithm. Both models achieve about 50% accuracy in the simulations but the LSTM model has the capability to acknowledge the market dependencies in the long term period. Jourdan *et al.* [60] formulated a few conditional dependencies induced by the block design of the Bitcoin protocol, and propose a probabilistic graphical model to predict the value of UTXOs, which record the number of Bitcoins in each transaction.

Another direction in this research area is to improve the mining strategies and power efficiencies using machine learning approaches. Wang *et al.* [61] employed Reinforcement Learning algorithm to dynamically analyze the profits of different mining strategies and discover the optimal mining strategies over time-varying blockchain networks. Some researchers demonstrate that the Bitcoin Mining can be quantified as a Markov Decision Process (MDP), and different reinforcement learning algorithms can be applied to construct the MDP model [62, 63]. Other than that, Nguyen *et al.* [64] present a reinforcement learning-based offloading scheme that assists mobile miners to determine optimal offloading decisions, reduce energy consumption, and avoid network latency.

4.1.3 Transaction entity classification

The Bitcoin has become an alternative medium of value exchange. Behind the screens, some users have taken advantage of the Bitcoin network for their illegal purposes. With the CoinJoin mixing service, Bitcoin has been recognized as a safe currency in the dark net markets and it can also be used for money laundering. In this case, there is an urgent need to develop transaction and address tracing systems. Machine learning has been considered as a powerful tool to perform cryptocurrency address clustering and labeling for detecting illegal activities.

In 2017, Yin and Vatrapu [65] built several different classifiers using supervised machine learning models, to identify the Bitcoin addresses that are related to criminal activities. The next year, Harlev *et al.* [66] also introduced a supervised learning model with the gradient boosting algorithm. All those learning models can achieve accuracy of 75% in the simulation of the address clustering. Besides that, Akcora *et al.* [67] proposed an efficient and tractable framework called BitcoinHeist. By applying topological data analysis into the past records of transactions, BitcoinHeist can automate the prediction of new ransomware transactions in an address cluster, and detect new ransomware that has no past records.

4.2 Blockchain-enabled machine learning model

While machine learning systems have become powerful tools to solve real-world problems, people started to question its trustworthiness. First of all, machine learning systems might be susceptible to data poisoning attacks. The hackers might try to manipulate the system performance by altering the collected data or inserting constructed poison instances. Secondly, it is difficult for humans to understand decisions made by the machine learning systems if there is no traceable logs or training histories. Thirdly, centralized servers are still heavily required for completing the model training processes. Finally, the model construction stages are not automated and the human involvement may bring biases into the final system. Blockchains and smart contracts have shown great potential to solve those challenges.

4.2.1 Blockchain for data security

Blockchain is known for keeping data secure and safe. With reliable and trace-able data stored on the blockchain, the researchers can ensure that machine learning algorithms will produce the most trusted and credible results. Muhammad *et al.* [68] introduced a federated learning system called Biscotti. Biscotti utilizes blockchain and cryptographic primitives to coordinate a privacy-preserving feder-ated learning process between peering clients. While all the training iterations are stored in the blockchain, only the peer-verified updates are committed into the final model. The training data are stored with the data providers locally. This system is able to protect the privacy of an individual client's data as well as defend data poisoning attacks.

Mugunthan *et al.* [69] provided a privacy-preserving federated learning system called BlockFlow. The system incorporates differential privacy, introduces a novel auditing mechanism for model contribution, and uses smart contracts to incentivize positive behaviors. However, the system does not have the capability to detect any anomalies during the learning process. To address that issue, Desai *et al.* [70] came up with another blockchain-based federated learning framework called BlockFLA. After the learning algorithm is deployed, the BlockFLA framework utilizes smart contracts to automatically detect and discourage any backdoor attacks by holding the responsible parties accountable. Both frameworks ensure that the machine learning algorithms are resilient to malicious adversaries.

In 2018, Chen *et al.* [71] proposed a secure supervised machine learning system called LearningChain. In LearningChain, they developed a differential privacy mechanism for the local gradient computing process to protect the privacy of individual data providers, and a l-nearest aggregation scheme to defend against Byzantine attacks in the global gradient's aggregation process. In the next year, Kim *et al.* [72] pointed out that the LearningChain system has several limitations such as low computation efficiency, zero support on non-deterministic function computa-tions, and weak privacy preservation. To revolve those issues in a systematic way, they developed an improved distributed machine learning model for permissioned blockchains. With a differentially private stochastic gradient descent method and an error-based aggregation rule as core primitives, their model provides better defences on the byzantine attacks and has the capability to handle the learning algorithm with non-deterministic functions defined. Besides that, Zhou [73] also introduced a similar system called PIRATE to provide distributed machine learning algorithms with byzantine-resiliency but the system is designed for 5G networks.

4.2.2 Blockchain for system improvement

Blockchains and smart contracts can also be utilized to improve the machine learning processes and eliminate human involvements. Ouyang *et al.* [74] implemented a novel federated learning collaboration framework: Learning Markets. In the Learning Markets, blockchain creates a trustless environment for collaboration and transaction. The learning task provider simply needs to publish the initial task to the markets and deposit the rewards in the network. The data providers and trainers participate in the learning process by depositing an entrance fee, uploading/downloading the data on IPFS network, and contributing their computation power. Several predefined smart contracts serve as network agents to maintain the collaboration relationships and market mechanisms.

Kim *et al.* [75] proposed an on-device blockchain-based Federated learning architecture called BlockFL. Data on user devices are processed locally and the local updates are accumulated on the blockchain. The global model updates are calculated based on the user updates recorded on each block. Their architecture mainly focuses on the latency minimization and system scalability. They also indicate that the system may not be able to retrieve all the local model updates on time due to network delay or intermittent availability problems.

Muhammad *et al.* [76] gave a complete list of requirements for the blockchain enabled federated learning framework, including penalisation, decentralization, fine-grained Federated Learning, incentive mechanisms, trust, activity monitoring, heterogeneity and context-awareness, model synchronization, and communication and bandwidth-efficiency. They also introduce a term called reputation (which is similar to the Proof of Stake) and describe how this attribute works in their proposed framework.

Besides that, some researchers in this subsection work on designing new blockchain mechanisms for the distributed machine learning tasks. Felipe *et al.* [77] invented a new protocol called Proof-of-Learning, which achieves distributed consensus by ranking machine learning systems for a given machine learning task. The aim of this protocol is to mitigate the computational consumption in solving hashing-based puzzles and still ensure the data integrity. Toyoda *et al.* [78] improved the common incentive mechanism in the current blockchain network and make it more applicable to the blockchain network when the machine learning tasks are involved.

4.3 Combined research for real world scenarios

Instead of proposing innovative and theoretical designs, some researchers intend to figure out how this combined research can be applied on some real-world problems. Their contribution establishes a bridge between this new research area and the traditional industries such as transportation, hospital management, supply chain, etc.

4.3.1 Transportation

Pokhrel and Choi [79] developed a mathematical framework that adapts the blockchain-based federated learning design into the autonomous vehicle industry. They utilize the consensus mechanism and a renewal reward approach to enable on-vehicle machine learning training in the distributed network. The on-vehicle updates and global models are maintained in the blockchain, which are visible to and verifiable by every vehicle. Rewards are distributed to the vehicle owners based on the size of their updates accepted into the global model. They also discussed the limitations of this design and the performance of the system based on the simulations and numerical analysis.

Hua *et al.* [80] tried to apply federated learning algorithm into the heavy haul railway management. In their research, the train controls are quantified into multiple classes and the individual train data is applicable to the SVM based mixed kernel. The global model is administered by the smart contract. This research resolves the data island issue in this industry and the asynchronous collaborative learning algorithm is designed without involving a central server.

4.3.2 Healthcare

The healthcare industry has long been an early adopter of and benefited greatly from technological advances. Chen et al. in [81] proposed a blockchain based disease-classification framework called Health-Chain. In the Health-Chain system, multiple institutes can train the model with their patient records, collaborate asynchronously in the blockchain network and contribute to the global model with privacy preserved. The researchers implement the system in two disease recognition tasks, breast cancer diagnosis and ECG arrhythmia classification, and both simulations demonstrate promising results.

Kumar *et al.* [82] proposed a similar but more elaborate supervised machine learning framework on detecting COVID-19 patients. The proposed framework can utilize up-to-date data which improves the recognition of CT images. Both researchers above mainly focus on building the machine learning models and the blockchain is used for enforcing the consensuses across research institutes and aggregating the training models.

Rahman *et al.* [83] gave a complete picture of how the blockchain can be employed in the Internet of Health Things (IoHT) area. In their framework, smart contracts are used to manage the training plan, trust management, participant authentication and the device data encryption. The framework design has high security and scalability level in the IoHT health management area.

4.3.3 Supply chain

Kamble *et al.* [84] built a prediction model using the machine learning technique to calculate an organization's probability of successful blockchain adoption (BA) within the supply chain industry. The researchers focus on explaining the intent of BA by using the psychological constructs from the literature on technology adoption. The learning model can help managers to predict the readiness of their organizations.

Mao *et al.* [85] developed a blockchain-based credit evaluation system to strengthen the efficiency of supervision and management in the food supply chain. The system collects credit evaluation from the traders on the blockchain, analyzes the evaluation directly via a deep learning network, and provides the credit results for the supervision and management of regulators.

Yong *et al.* [86] proposed a detailed" vaccine blockchain" system based on blockchain and machine learning technologies. The vaccine system is designed to support tracing the vaccine inventory and preventing supply record fraud. The machine learning model can also provide suggestions to the immunization practitioners and recipients.

5. Discussions and conclusion

5.1 Current threads in Blockchain research

Blockchain systems are designed to be distributed. Ironically, most blockchain networks are facing problems on centralization. For example, PoW mechanism relies on the success of mining mechanism. The higher the hashrate is, the more resilient the Bitcoin is against attacks. However, on the market side, mining a Bitcoin becomes increasingly harder. In order to gain more reward, miners share their hash power under a common mining pool. As shown in **Figure 5**, the top 4 mining pools currently contribute to more than 50% of the entire hash power.

In other words, if a hacker controls or manipulates the top4 mining pools, they might be able to perform 51% of the attacks. This is known as mining pool attacks. Besides directly controlling the mining pool, multiple variations of mining pool attacks have also been proposed. Selfish mining, for instance, refers to when miners who find the next block withhold the information and then release multiple valid blocks at once, resulting in other miners losing their block rewards and blocks.

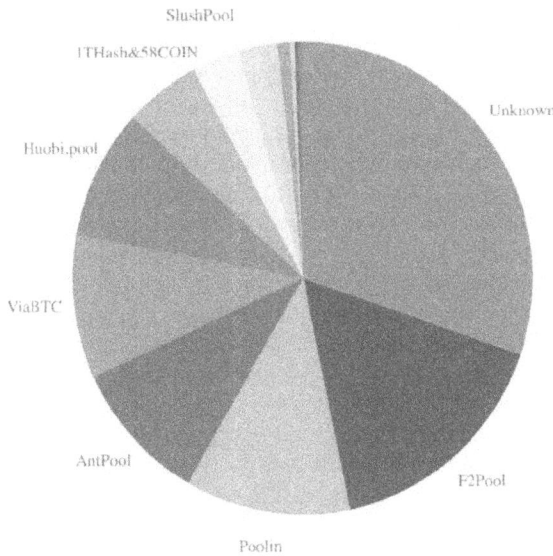

Figure 5.
Bitcoin Main Network Hashrate Distribution [87].

Many upgraded versions of mining pool attacks have been brought forth, such as Fork-after-Withhold (FAW) [88].

Ideally speaking, blockchain is resilient to Denial-of-Service Attack (DDoS) due to its distributed characteristics. However, the network layer is not completely decentralized. It includes routers, Internet Service Provider (ISP) for mining pools and cloud services. More than 60% of the Bitcoin nodes are hosted in less than 100 IP prefixes [89]. Such attacks on network layers are easier to perform and could have a larger impact on the entire blockchain network. *Hijacking Bitcoin: Routing Attacks on Cryptocurrencies* by Apostolaki *et al.* had talked about network layer attacks, partitioning attacks, and delay attacks [89].

5.2 Quantum computing

The existence and advancement of quantum computers will bring a massive change to our current technology industry, from cryptography [90], artificial intelligence [91] to computer architecture [92]. Nowadays, small to intermediate-scale quantum computers already exist in universities and industry laboratories (Noisy intermediate Scale quantum devices, often called NISQ [93]). Such noisy devices with about 50 qubits are promising to demonstrate quantum supremacy in the following years [94]. Quantum computers are devices using quantum phenomena such as superposition and entanglement to perform calculation. Quantum computers are believed to solve certain computational problems, such as integer factorization, substantially faster than traditional computers [94].

Blockchain security heavily relies on asymmetric encryption [90]. Besides, hash functions are commonly used by most blockchain networks in order to compress the content of all information. Both hash functions and asymmetric encryption are threatened by the evolution of quantum computers due to shor's algorithm [90], a polynomial-time solution to integer factorization problems invented by Peter Shor in 1994 [95]. Asymmetric encryption or public key encryption was designed based on elliptic curve cryptography (ECC) [90]. Quantum computers dominate ECDSA

(Elliptic Curve Digital Signature Algorithm), a secure and efficient tool used in Bitcoin systems [7]. Hence, a quantum resilient and high efficiency algorithm is needed for future Bitcoin/blockchain development [90].

Another algorithm that will bring a huge impact on blockchain is called Grover's algorithm. Grover's algorithm is a quantum computing algorithm that can quadratically speed up the unstructured search problem [96]. Furthermore, this algorithm has been used as a general trick or subroutine to obtain quadratic run time improvements for many other algorithms [97]. Firstly, grover's algorithm can be used to find collisions in hash functions, causing hash function to lose security. Secondly, Grover's algorithm can be used to accelerate mining because of its efficiency in searching for nonces, resulting in biasing in computational power and further recreating entire blockchains to manipulate the historical transactions [90].

Last but not least, to successfully implement Shor's algorithm, it will require more than 5,000 qubits to factor cryptographically significant numbers [98]. That is only without considering the error correction properties of quantum mechanisms. With error correction, the requirement may go up to as high as a million. In addition to the large number of qubits, it also requires hundreds of millions of gate operations [98]. This requirement is nearly impossible to achieve today, as Google's best quantum computer can only reach 54 qubits in 2019. Note that there are many approaches to build quantum computers; the qubits numbers referred to here is based on digitized adiabatic quantum computers with a superconducting circuit [99]. Another famous approach is using quantum annealing [100], led by D-Wave.

5.3 Conclusion

In this part, we reviewed and summarized the state-of-the-art research papers related to combining blockchain and AI in different scenarios. We provided a survey about how blockchain network could be integrated into the MEC technology. With security and privacy-preserving features, blockchain could be an effective solution to secure the aspects of data sharing and resources allocation in AI applications, especially for mobile edge intelligence. Besides, we introduced multiple research papers with respect to how machine learning and blockchain collaborate with each other. While machine learning can be utilized to improve network security and stability of blockchain, blockchain can also automate the model learning process and protect sensitive training data. We further discussed some threats that would challenge blockchain systems, including malicious attacks and quantum computing. Overall, this chapter demonstrates that Blockchain and AI researches are still at an early stage. Once all the bottlenecks and challenges in this combined research area are addressed, blockchain network could become a necessary and important platform to enable and improve AI applications across different industries.

Acknowledgements

This work was supported by Blockchain@UBC and Natural Sciences and Engineering Research Council of Canada (CREATE Program grant 528125).

Blockchain-Empowered Mobile Edge Intelligence, Machine Learning and Secure Data Sharing
DOI: *http://dx.doi.org/10.5772/intechopen.96618*

Author details

Yao Du[†], Shuxiao Miao[†], Zitian Tong[†], Victoria Lemieux and Zehua Wang[*]
Department of Electrical and Computer Engineering, The University of British
Columbia, Vancouver, BC, Canada

[*]Address all correspondence to: zwang@ece.ubc.ca

[†] These authors contributed equally.

IntechOpen

References

[1] Jameel F, Javaid U, Khan WU, Aman MN, Pervaiz H, Jantti R. Reinforcement Learning in Blockchain-Enabled IIoT Networks: A Survey of Recent Advances and Open Challenges. Sustainability. 2020;12(12):5161.

[2] Patel M, Naughton B, Chan C, Sprecher N, Abeta S, Neal A, et al. Mobile-edge computing introductory technical white paper. White paper, mobile-edge computing (MEC) industry initiative. 2014;29:854–864.

[3] Nguyen DC, Pathirana PN, Ding M, Seneviratne A. Privacy-Preserved Task Offloading in Mobile Blockchain With Deep Reinforcement Learning. IEEE Transactions on Network and Service Management. 2020;17(4):2536–2549.

[4] Xiong Z, Zhang Y, Niyato D, Wang P, Han Z. When Mobile Blockchain Meets Edge Computing. IEEE Communications Magazine. 2018; 56(8):33–39.

[5] Conti M, Sandeep Kumar E, Lal C, Ruj S. A Survey on Security and Privacy Issues of Bitcoin. IEEE Communications Surveys Tutorials. 2018;20(4):3416–3452.

[6] Worrall E. Study: Bitcoin Mining Could Push Global Warming Over the 2C Threshold; 2018. Copyright - Copyright Newstex Oct 29, 2018; Last updated - 2019-07-08.

[7] Nakamoto S. Bitcoin: A peer-to-peer electronic cash system. bitcoin.org; 2008.

[8] Ferdous MS, Chowdhury MJM, Hoque MA, Colman A. Blockchain Consensus Algorithms: A Survey; 2020.

[9] Tseng L. Recent Results on Fault-Tolerant Consensus in Message-Passing Networks; 2016.

[10] Lamport L, Shostak R, Pease M. The Byzantine Generals Problem. ACM transactions on programming languages and systems. 1982;4(3):382–401.

[11] Gupta KD, Rahman A, Poudyal S, Huda MN, Mahmud MAP. A Hybrid POW-POS Implementation Against 51 percent Attack in Cryptocurrency System. In: 2019 IEEE International Conference on Cloud Computing Technology and Science (CloudCom); 2019. p. 396–403.

[12] Küfeoğlu S, Özkuran M. Bitcoin mining: A global review of energy and power demand. Energy research social science. 2019;58:101273.

[13] Wang YZ, Wu J, Chen SH, Chao MC, Yang CH. Micro-Architecture Optimization for Low-Power Bitcoin Mining ASICs. IEEE; 2019. p. 1–4.

[14] Saleh F. Blockchain without Waste: Proof-of-Stake. The Review of financial studies. 2020.

[15] King S, Nadal S. PPCoin: Peer-to-Peer Crypto-Currency with Proof-of-Stake; 2012.

[16] Lepore C, Ceria M, Visconti A, Rao UP, Shah KA, Zanolini L. A Survey on Blockchain Consensus with a Performance Comparison of PoW, PoS and Pure PoS. Mathematics (Basel). 2020;8(1782):1782.

[17] Reijsbergen D, Szalachowski P, Ke J, Li Z, Zhou J. LaKSA: A Probabilistic Proof-of-Stake Protocol; 2021.

[18] Mistry I, Tanwar S, Tyagi S, Kumar N. Blockchain for 5G-enabled IoT for industrial automation: A systematic review, solutions, and challenges. Mechanical systems and signal processing. 2020;135:106382.

[19] Yazdinejad A, Srivastava G, Parizi RM, Dehghantanha A, Karimipour H, Karizno SR. SLPoW: Secure and Low Latency Proof of Work Protocol for Blockchain in Green IoT Networks. In: 2020 IEEE 91st Vehicular Technology Conference (VTC2020-Spring); 2020. p. 1–5.

[20] Varga P, Peto J, Franko A, Balla D, Haja D, Janky F, et al. 5G support for Industrial IoT Applications - Challenges, Solutions, and Research gaps. Sensors (Basel, Switzerland). 2020;20(3):828.

[21] Fan K, Ren Y, Wang Y, Li H, Yang Y. Blockchain-based efficient privacy preserving and data sharing scheme of content-centric network in 5G. IET communications. 2018;12(5): 527–532.

[22] in Media BBSL, Communications, Lecturer DB. Netflix's The Social Dilemma highlights the problem with social media, but what's the solution?; 2020. Available from: https://theconve rsation.com.

[23] Podgorelec B, Kersic V, Turkanovic M. Analysis of Fault Tolerance in Permissioned Blockchain Networks. IEEE; 2019. p. 1–6.

[24] Goldwasser S, Micali S, Rackoff C. The Knowledge Complexity of Interactive Proof Systems. SIAM Journal on Computing. 1989 02;18(1):186–23. Copyright - Copyright] © 1989 Society for Industrial and Applied Mathematics; Last updated - 2012-02-05.

[25] D RR, Adam S, Katerina S. Toward Non-interactive Zero-Knowledge Proofs for NP from LWE. Journal of cryptology. 2021;34(1).

[26] Kumar A, Fischer C, Tople S, Saxena P. In: A Traceability Analysis of Monero's Blockchain. Cham: Springer International Publishing; 2017. p. 153–173.

[27] hour ago Major Russian Bank Sberbank Files Application to Launch Its Own Stablecoin — Possibly Pegged to the Fiat Ruble ALTCOINS — 20 hours ago PCMASEWSBMPA, to be a Digital Bank in Gibraltar Biden Administration Reported to Be Lining up a Former Ripple Advisor as the Next Bank Regulator Bitcoin Near 'Extreme Bubble' but Tesla More Vulnerable: Deutsche Bank Survey Russia Prohibits Government Officials From Owning Crypto CCXS. Not So Private: 99% of Zcash and Dash Transactions Traceable, Says Chainalysis – Altcoins Bitcoin News; 2020. Available from: https://ne ws.bitcoin.com/.

[28] Abbas N, Zhang Y, Taherkordi A, Skeie T. Mobile Edge Computing: A Survey. IEEE Internet of Things Journal. 2018 Feb;5(1):450–465.

[29] Xiao L, Ding Y, Jiang D, Huang J, Wang D, Li J, et al. A Reinforcement Learning and Blockchain-Based Trust Mechanism for Edge Networks. IEEE Transactions on Communications. 2020; 68(9):5460–5470.

[30] Seng S, Li X, Luo C, Ji H, Zhang H. A D2D-assisted MEC Computation Offloading in the Blockchain-Based Framework for UDNs. In: ICC 2019–2019 IEEE International Conference on Communications (ICC). New York: IEEE; 2019. .

[31] Liu M, Yu FR, Teng Y, Leung VCM, Song M. Joint Computation Offloading and Content Caching for Wireless Blockchain Networks. In: IEEE Infocom 2018 - IEEE Conference on Computer Communications Workshops (infocom Wkshps). New York: IEEE; 2018. p. 517–522.

[32] Liu M, Yu FR, Teng Y, Leung VCM, Song M. Computation Offloading and Content Caching in Wireless Blockchain Networks With Mobile Edge Computing. IEEE Transactions on

Vehicular Technology. 2018;67(11): 11008–11021.

[33] Liu M, Yu FR, Teng Y, Leung VCM, Song M. Distributed Resource Allocation in Blockchain-Based Video Streaming Systems With Mobile Edge Computing. IEEE Transactions on Wireless Communications. 2019;18(1): 695–708.

[34] Feng J, Yu FR, Pei Q, Chu X, Du J, Zhu L. Cooperative Computation Offloading and Resource Allocation for Blockchain-Enabled Mobile-Edge Computing: A Deep Reinforcement Learning Approach. IEEE Internet of Things Journal. 2020;7(7):6214–6228.

[35] Zhang Z, Hong Z, Chen W, Zheng Z, Chen X. Joint Computation Offloading and Coin Loaning for Blockchain-Empowered Mobile-Edge Computing. IEEE Internet of Things Journal. 2019;6(6):9934–9950.

[36] Liu W, Cao B, Zhang L, Peng M, Daneshmand M. A Distributed Game Theoretic Approach for Blockchain-based Offloading Strategy. In: ICC 2020–2020 IEEE International Conference on Communications (ICC); 2020. p. 1–6.

[37] Hassija V, Chamola V, Gupta V, Chalapathi GSS. A Blockchain based Framework for Secure Data Offloading in Tactile Internet Environment. In: 2020 International Wireless Communications and Mobile Computing (IWCMC); 2020. p. 1836–1841.

[38] Zhang S, Lee J. A Group Signature and Authentication Scheme for Blockchain-Based Mobile-Edge Computing. IEEE Internet of Things Journal. 2020 May;7(5):4557–4565.

[39] Rahman MA, Hossain MS, Islam MS, Alrajeh NA, Muhammad G. Secure and Provenance Enhanced Internet of Health Things Framework: A Blockchain Managed Federated

Learning Approach. IEEE ACCESS. 2020;8:205071–205087.

[40] Kang J, Yu R, Huang X, Wu M, Maharjan S, Xie S, et al. Blockchain for Secure and Efficient Data Sharing in Vehicular Edge Computing and Networks. IEEE Internet of Things Journal. 2019 Jun;6(3):4660–4670.

[41] Lu Y, Huang X, Dai Y, Maharjan S, Zhang Y. Blockchain and federated learning for privacy-preserved data sharing in industrial IoT. IEEE Transactions on Industrial Informatics. 2019;16(6):4177–4186.

[42] Islam A, Shin SY. BUAV: A Blockchain Based Secure UAV-Assisted Data Acquisition Scheme in Internet of Things. Journal of Communications and Networks. 2019;21(5):491–502.

[43] Zhang W, Lu Q, Yu Q, Li Z, Liu Y, Lo SK, et al. Blockchain-based Federated Learning for Device Failure Detection in Industrial IoT. IEEE Internet of Things Journal. 2020.

[44] Zyskind G, Nathan O, Pentland A. Decentralizing Privacy: Using Blockchain to Protect Personal Data. In: 2015 IEEE Security and Privacy Workshops; 2015. p. 180–184.

[45] Yang H, Liang Y, Yuan J, Yao Q, Yu A, Zhang J. Distributed Blockchain-Based Trusted Multidomain Collaboration for Mobile Edge Computing in 5G and Beyond. IEEE Transactions on Industrial Informatics. 2020;16(11):7094–7104.

[46] Arachchige PCM, Bertok P, Khalil I, Liu D, Camtepe S, Atiquzzaman M. A Trustworthy Privacy Preserving Framework for Machine Learning in Industrial IoT Systems. IEEE Transactions on Industrial Informatics. 2020 Sep;16(9):6092–6102.

[47] Lin X, Wu J, Bashir AK, Li J, Yang W, Piran J. Blockchain-Based

Incentive Energy-Knowledge Trading in IoT: Joint Power Transfer and AI Design. IEEE Internet of Things Journal. 2020:1–14.

[48] Fan S, Zhang H, Zeng Y, Cai W. Hybrid Blockchain-Based Resource Trading System for Federated Learning in Edge Computing. IEEE Internet of Things Journal. 2020.

[49] Liu Y, Yu FR, Li X, Ji H, Leung VCM. Resource Allocation for Video Transcoding and Delivery Based on Mobile Edge Computing and Blockchain. In: 2018 IEEE Global Communications Conference (GLOBECOM); 2018. p. 1–6.

[50] Zhang R, Yu FR, Liu J, Huang T, Liu Y. Deep Reinforcement Learning (DRL)-Based Device-to-Device (D2D) Caching With Blockchain and Mobile Edge Computing. IEEE Transactions on Wireless Communications. 2020;19(10): 6469–6485.

[51] Somy NB, Kannan K, Arya V, Hans S, Singh A, Lohia P, et al. Ownership Preserving AI Market Places Using Blockchain. In: 2019 IEEE International Conference on Blockchain (Blockchain); 2019. p. 156–165.

[52] Chen H, Pendleton M, Njilla L, Xu S. A Survey on Ethereum Systems Security: Vulnerabilities, Attacks, and Defenses. ACM Comput Surv. 2020 Jun;53(3).

[53] Scicchitano F, Liguori A, Guarascio M, Ritacco E, Manco G. A Deep Learning Approach for Detecting Security Attacks on Blockchain; 2020. .

[54] Dey S. Securing Majority-Attack in Blockchain Using Machine Learning and Algorithmic Game Theory: A Proof of Work. In: 2018 10th Computer Science and Electronic Engineering (CEEC); 2018. p. 7–10.

[55] Hou C, Zhou M, Ji Y, Daian P, Tramer F, Fanti G, et al.. SquirRL:

Automating Attack Analysis on Blockchain Incentive Mechanisms with Deep Reinforcement Learning; 2020.

[56] Kim YB, Kim JG, Kim W, Im JH, Kim TH, Kang SJ, et al. Predicting Fluctuations in Cryptocurrency Transactions Based on User Comments and Replies. PLOS ONE. 2016 08;11(8): 1–17.

[57] Madan I. Automated Bitcoin Trading via Machine Learning Algorithms; 2014. .

[58] Jang H, Lee J. An Empirical Study on Modeling and Prediction of Bitcoin Prices With Bayesian Neural Networks Based on Blockchain Information. IEEE Access. 2018;6:5427–5437.

[59] McNally S, Roche J, Caton S. Predicting the Price of Bitcoin Using Machine Learning. In: 2018 26th Euromicro International Conference on Parallel, Distributed and Network-based Processing (PDP); 2018. p. 339–343.

[60] Jourdan M, Blandin S, Wynter L, Deshpande P. A Probabilistic Model of the Bitcoin Blockchain. In: 2019 IEEE/ CVF Conference on Computer Vision and Pattern Recognition Workshops (CVPRW); 2019. p. 2784–2792.

[61] Wang T, Liew SC, Zhang S. When Blockchain Meets AI: Optimal Mining Strategy Achieved By Machine Learning. CoRR. 2019;abs/1911.12942.

[62] Eyal I, Sirer EG. Majority is not Enough: Bitcoin Mining is Vulnerable. CoRR. 2013;abs/1311.0243. Available from: http://arxiv.org/abs/1311.0243.

[63] Sapirshtein A, Sompolinsky Y, Zohar A. Optimal Selfish Mining Strategies in Bitcoin. CoRR. 2015;abs/ 1507.06183. Available from: http://arxiv. org/abs/1507.06183.

[64] Nguyen DC, Pathirana PN, Ding M, Seneviratne A. Privacy-Preserved Task

Offloading in Mobile Blockchain With Deep Reinforcement Learning. IEEE Transactions on Network and Service Management. 2020 Dec;17(4):2536–2549.

[65] Sun Yin H, Vatrapu R. A first estimation of the proportion of cybercriminal entities in the bitcoin ecosystem using supervised machine learning. In: 2017 IEEE International Conference on Big Data (Big Data); 2017. p. 3690–3699.

[66] Harlev MA, Yin H, Langenheldt KC, Mukkamala R, Vatrapu R. Breaking Bad: De-Anonymising Entity Types on the Bitcoin Blockchain Using Supervised Machine Learning. In: HICSS; 2018.

[67] Akcora CG, Li Y, Gel YR, Kantarcioglu M. BitcoinHeist: Topological Data Analysis for Ransomware Prediction on the Bitcoin Blockchain. In: Bessiere C, editor. Proceedings of the Twenty-Ninth International Joint Conference on Artificial Intelligence, IJCAI-20. International Joint Conferences on Artificial Intelligence Organization; 2020. p. 4439–4445. Special Track on AI in FinTech.

[68] Shayan M, Fung C, Yoon CJM, Beschastnikh I. Biscotti: A Ledger for Private and Secure Peer-to-Peer Machine Learning. CoRR. 2018;abs/1811.09904. Available from: http://arxiv.org/abs/1811.09904.

[69] Mugunthan V, Rahman R, Kagal L. BlockFLow: An Accountable and Privacy-Preserving Solution for Federated Learning; 2020.

[70] Desai HB, Ozdayi MS, Kantarcioglu M. BlockFLA: Accountable Federated Learning via Hybrid Blockchain Architecture; 2020.

[71] Chen X, Ji J, Luo C, Liao W, Li P. When Machine Learning Meets Blockchain: A Decentralized, Privacy-preserving and Secure Design. In: 2018 IEEE International Conference on Big Data (Big Data); 2018. p. 1178–1187.

[72] Kim H, Kim S, Hwang JY, Seo C. Efficient Privacy-Preserving Machine Learning for Blockchain Network. IEEE Access. 2019;7:136481–136495.

[73] Zhou S, Huang H, Chen W, Zheng Z, Guo S. PIRATE: A Blockchain-based Secure Framework of Distributed Machine Learning in 5G Networks. CoRR. 2019;abs/1912.07860. Available from: http://arxiv.org/abs/1912.07860.

[74] Ouyang L, Yuan Y, Wang FY. Learning Markets: An AI Collaboration Framework Based on Blockchain and Smart Contracts. IEEE Internet of Things Journal. 2020.

[75] Kim H, Park J, Bennis M, Kim S. Blockchained On-Device Federated Learning. IEEE Communications Letters. 2020;24(6):1279–1283.

[76] ur Rehman MH, Salah K, Damiani E, Svetinovic D. Towards Blockchain-Based Reputation-Aware Federated Learning. In: IEEE INFOCOM 2020 - IEEE Conference on Computer Communications Workshops (INFOCOM WKSHPS); 2020. p. 183–188.

[77] Bravo-Marquez F, Reeves S, Ugarte M. Proof-of-Learning: A Blockchain Consensus Mechanism Based on Machine Learning Competitions. In: 2019 IEEE International Conference on Decentralized Applications and Infrastructures (DAPPCON); 2019. p. 119–124.

[78] Toyoda K, Zhang AN. Mechanism Design for An Incentive-aware Blockchain-enabled Federated Learning Platform. In: 2019 IEEE International Conference on Big Data (Big Data); 2019. p. 395–403.

[79] Pokhrel SR, Choi J. Federated Learning With Blockchain for

Autonomous Vehicles: Analysis and
Design Challenges. IEEE Transactions on
Communications. 2020;68(8):4734–4746.

[80] Hua G, Zhu L, Wu J, Shen C,
Zhou L, Lin Q. Blockchain-Based
Federated Learning for Intelligent
Control in Heavy Haul Railway. IEEE
Access. 2020;8:176830–176839.

[81] Chen X, Wang X, Yang K.
Asynchronous Blockchain-based
Privacy-preserving Training Framework
for Disease Diagnosis. In: 2019 IEEE
International Conference on Big Data
(Big Data); 2019. p. 5469–5473.

[82] Kumar R, Khan AA, Zhang S,
Kumar J, Yang T, Golalirz NA, et al..
Blockchain-Federated-Learning and
Deep Learning Models for COVID-19
detection using CT Imaging; 2020.

[83] Rahman MA, Hossain MS,
Islam MS, Alrajeh NA, Muhammad G.
Secure and Provenance Enhanced
Internet of Health Things Framework: A
Blockchain Managed Federated
Learning Approach. IEEE Access. 2020;
8:205071–205087.

[84] Kamble S, Gunasekaran A,
Kumar V, Belhadi A, Foropon C. A
machine learning based approach for
predicting blockchain adoption in
supply Chain. Technological Forecasting
and Social Change. 2020 11.

[85] Mao D, Wang F, Hao Z, Li H. Credit
Evaluation System Based on Blockchain
for Multiple Stakeholders in the Food
Supply Chain. International Journal of
Environmental Research and Public
Health. 2018 08;15:1627.

[86] Yong B, Shen J, Liu X, Li F, Chen H,
Zhou Q. An intelligent blockchain-based
system for safe vaccine supply and
supervision. International Journal of
Information Management. 2020;52:
102024. Available from: http://www.
sciencedirect.com/science/article/pii/
S0268401219304505.

[87] blockchain com. Hashrate
Distribution: An estimation of hashrate
distribution amongst the largest mining
pools; 2021. https://www.blockchain.
com/pools.

[88] Kwon Y, Kim D, Son Y,
Vasserman E, Kim Y. Be Selfish and
Avoid Dilemmas: Fork After
Withholding (FAW) Attacks on Bitcoin.
2017.

[89] Apostolaki M, Zohar A, Vanbever L.
Hijacking Bitcoin: Routing Attacks on
Cryptocurrencies. IEEE; 2017.
p. 375–392.

[90] Fernández-Caramès TM, Fraga-
Lamas P. Towards Post-Quantum
Blockchain: A Review on Blockchain
Cryptography Resistant to Quantum
Computing Attacks. IEEE Access. 2020;
8:21091–21116.

[91] Choi J, Oh S, Kim J. The Useful
Quantum Computing Techniques for
Artificial Intelligence Engineers. In:
2020 International Conference on
Information Networking (ICOIN);
2020. p. 1–3.

[92] Riesebos L, Fu X, Moueddenne AA,
Lao L, Varsamopoulos S, Ashraf I, et al.
Quantum Accelerated Computer
Architectures. In: 2019 IEEE
International Symposium on Circuits
and Systems (ISCAS); 2019. p. 1–4.

[93] Tanimoto T, Matsuo S, Kawakami S,
Tabuchi Y, Hirokawa M, Inoue K. How
Many Trials Do We Need for Reliable
NISQ Computing? In: 2020 IEEE
Computer Society Annual Symposium
on VLSI (ISVLSI); 2020. p. 288–290.

[94] Arute F, Arya K, Babbush R,
Bacon D, Bardin J, Barends R, et al.
Quantum supremacy using a
programmable superconducting
processor. Nature. 2019 10;574:505–510.

[95] Shor PW. Polynomial-Time
Algorithms for Prime Factorization and

Discrete Logarithms on a Quantum Computer. SIAM Journal on Computing. 1997 Oct;26(5):1484–1509.

[96] Grover L. Fast quantum mechanical algorithm for database search. Proceedings of the 28th Annual ACM Symposium on the Theory of Computing. 1996 06.

[97] Team TQ. Grover's Algorithm. Data 100 at UC Berkeley; 2021. Available from: https://qiskit.org/textbook/ch-algorithms/grover.html.

[98] Guerreschi GG, Matsuura AY. QAOA for Max-Cut requires hundreds of qubits for quantum speed-up. Scientific Reports. 2019 May;9(1).

[99] Barends R, Shabani A, Lamata L, Kelly J, Mezzacapo A, Heras UL, et al. Digitized adiabatic quantum computing with a superconducting circuit. Nature. 2016 Jun;534(7606):222–226.

[100] DE FALCO D, TAMASCELLI D. AN INTRODUCTION TO QUANTUM ANNEALING. vol. 45. Les Ulis: EDP Sciences; 2011. p. 99–116.

Chapter 5

Blockchain and AI Meet in the Metaverse

Hyun-joo Jeon, Ho-chang Youn, Sang-mi Ko and Tae-heon Kim

Abstract

With new technologies related to the development of computers, graphics, and hardware, the virtual world has become a reality. As COVID-19 spreads around the world, the demand for virtual reality increases, and the industry represented by the Metaverse is developing. In the Metaverse, a virtual world that transcends reality, artificial intelligence and blockchain technology are being combined. This chapter explains how artificial intelligence and blockchain can affect the Metaverse.

Keywords: AI, Blockchain, Virtual Reality, Metaverse, NFT

1. Introduction

The term "Metaverse" is a combination of 'meta' meaning 'virtual, transcendence' and 'verse' a backformation from 'universe'. The Acceleration Studies Foundation (ASF), a non-profit technology research organization, classified the Metaverse into the following four categories: a virtual world that experiences a flawless virtual story, a mirror world that reflects the current real world, an augmented reality that shows a mixture of augmented information in the real world and life logging, which captures and stores everyday information about people and things [1]. With the development of technology, the number of people who use the Metaverse increases, and as activities at the same level as reality are performed, various and a lot of data are being generated. Data generated in the metaverse has value in itself. In the Metaverse, the amount of data increases, the value increases, and the importance of reliability and security is increasing. Blockchain technology is required to guarantee the reliability of data in the Metaverse, and artificial intelligence is used to secure the diversity and rich content of the Metaverse. The contents will be developed in the following order.

In this chapter, under the theme of the Metaverse, we will look into the issues of human instinct for creation in the virtual world, the phenomenon in which the real and the virtual are combined in the virtual world, and the reliability of data in this virtual world. Blockchain and NFT technologies are described as trust technologies. And the Metaverse platform built on the basis of this technology will be described. Basically, we will understand the interface between blockchain and artificial intelligence, and look at how a better world is created by combining blockchain and artificial intelligence in Metaverse.

2. Virtual world and desire of creation

2.1 Human desire for creation

Humans have an instinct for creation, and this creativity is an important factor that distinguishes humans from other animals. The creativity of human beings has been creating the culture. The paper published in 2004 described the SeaCircle as the new concept of the culture, and it regarded the SeaCircle as human cultural activities for creating. In the concept of the SeaCircle, humans are the spiritual beings, and only humans constitute a culture. It explains the elements of insight of culture [2]. According to the SeaCircle theory, creativity is explained as an element of Open Mind and Spirit [3].

On the SeaCircle concept, the Metaverse can be interpreted as a space that allows people to be more immersed in creative activities by resolving some of the constraints on space and resources.

2.2 Connection between the virtual world and the real world

Recently, the virtual world and the real world have been developed in convergence. The First and Second Industrial Revolutions were the process of maximizing efficiency through division of labor, so the production of materials and the consumption of materials were separated. In the Third Industrial Revolution, as online transactions are actively conducted, data has become an important commodity, and offline transactions are gradually being replaced by online. In the Firth Industrial Revolution, an intelligent revolution is occurring as things and humans become hyper-connected. There is a convergence phenomenon in which production and consumption occur at the same time, such as social customization or digital DIY(Design It Yourself). The offline world composed of materials is dominated by Pareto's law, which attempts to own and concentrate on the core of 20% due to limited resources. On the other hand, in the online world of information, the Long Tail theory is applied to share and find opportunities from the marginalized 80% of customers. The Forth Industrial Revolution is creating a convergence world where offline and online meet. This convergence is being created in manufacturing, logistics, finance, automotive, sports, healthcare, education, food and everyday life. In addition, as the problems of material production and supply were solved in the First, Second, and the Third Industrial Revolutions, and interest in human personal desire and spirit increased in the Forth Industrial Revolution stage, a new convergence between offline world and online world are being created [4].

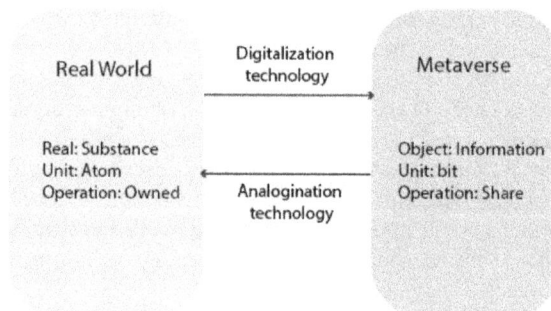

Figure 1.
Relationship between the real world and the Metaverse.

2.3 Combination of virtual and real in Metaverse

Political, economic, social, and cultural interactions appear in the Metaverse, which seems to mimic the real world. **Figure 1** shows the process of interworking and convergence between the real world and the Metaverse [4]. The Metaverse expresses an alternative world that cannot be achieved well in the real world.

In Minecraft, a virtual reality game platform representing Metaverse, as it became difficult to go to school due to COVID-19, UC Berkeley students created a campus inside the Minecraft game and held an event to hold a virtual graduation ceremony [5]. The president, guest speakers, and graduates all participated as Minecraft characters, and even the tradition of throwing hats after graduation was reenacted in Minecraft.

Roblox allows game developers to create games on the Roblox virtual platform instead of commuting to an offline office [6]. In Roblox, tokens are obtained in return for labor, and the tokens obtained in the game can be brought to the outside to be cashed.

In Metaverse, numerous users can freely trade goods and services according to the currency and transaction method provided by the platform. Both the virtual asset SAND of The Sandbox and the virtual asset MANA in Digital Land are listed on the exchange and are actively traded [7, 8]. This means that money in the virtual world in units of bits can replace money in the real world. Co-creating a game in the space of the virtual world means replacing the space in the space of the real world. This means that activities in the real world are data in bits in the virtual space, and the importance and reliability of these data are emphasized.

3. Data trust in the virtual world

After the 4th industrial revolution, the virtual world has grown rapidly. The real thing has been converted into data from the virtual world, and the virtual world has even played a role in leading the real world. Here, we have a question about the reliability of data about whether the real thing is becoming data accurately in the virtual world. In the virtual world, trust technology is emerging as an important issue. We can think of blockchain as one of these trust technologies.

Blockchain was first proposed in 2008 in Satoshi Nakamoto's paper "Bitcoin: P2P Electronic Money System". Blockchain can be said to be a technology that gives trust in transactions between individuals. A blockchain consists of blocks containing data and a chain that connects them. It is a blockchain to create and connect blocks, and consensus algorithms are used in this process. Any of the nodes participating in the network can create blocks, but not all generated blocks are connected, and only one block is recognized and connected. Since only one block among many blocks is connected to the previous block and the remaining blocks are discarded, consensus among participating nodes to select one block is essential. As a method of reaching consensus, consensus algorithms such as Proof of Work (PoW) and Proof of Stake (PoS) are used. If it is recognized as a unique block by all nodes, the node that created the block will receive cryptocurrency as a reward. This action is called mining, and a blockchain connected only with blocks created by mining is called a Canonical Chain.

Blockchain is developing and evolving as shown in **Table 1** [9]. Blockchain 1.0 was a period of innovation in the financial system with the advent of Bitcoin. Bitcoin is meaningful in that it attempts a single global financial system based on decentralization and decentralization, which are the core values of blockchain.

Blockchain 1.0	Blockchain 2.0	Blockchain 3.0
Crypto currency, Currency transfer, Remittance. Digital payment system	Smart contract, Decentralized autonomous organization(DAO), Stock, Bonds, Loans, Mortgages, Smart property	Government, Public, Science, Health, Culture, Art, IoT, Big Data, AI

Table 1.
Blockchain paradigm evolution direction.

Blockchain 2.0 is a period of contract automation centered on Ethereum smart contracts. It made it possible to execute contracts with legal effect online only with computer code without a transaction intermediary. It is a period that showed the potential for development as an online trading platform. Blockchain 3.0 is the stage in which blockchain technology is spread and applied to various industries. In order to solve the problems of the previous blockchain, technological improvements such as changes in consensus algorithms, improvement of transaction processing speed, and in-house decision-making functions are being made [10]. While it is expected that artificial intelligence will be applied to more expanded fields in Blockchain 3.0, more various applications of blockchain and artificial intelligence are expected to appear in the Metaverse environment.

4. Blockchain-based Metaverse

4.1 Ethereum code

Ethereum is a platform network designed to operate various decentralized applications (DApps), based on its own blockchain. Just as the basic framework and details of Internet standards were documented as RFCs, Ethereum Request for Comment (ERC) documents the details of Ethereum. In DApps using the Ethereum network, the basic protocol for issuing tokens is expressed as ERC-number as shown in **Table 2**. ERC-number is a protocol to follow when issuing tokens from DApps using the Ethereum network. Ethereum standard documents start with ERC-20, ERC-165, ERC-223, ERC-621, ERC-721, ERC-777, ERC-827, ERC-884, ERC-998, ERC-1155, ERC-1404 etc. [11].

Among them, ERC-20 is a protocol related to replaceable tokens, and ERC-20 tokens have the same value and function and can be exchanged with each other. The Ethereum project is issuing tokens based on ERC-20 and allowing investment and various businesses to take place.

ERC-721 is a protocol for NFT (Non-Fungible Tokens). NFT guarantees uniqueness by keeping encrypted transaction history permanently on the blockchain. Each token has a unique recognition value, authenticating the ownership of digital assets and assigning a value to the transaction. NFT has been mainly used to commemorate special moments or to collect digital assets, and recently it is creating a new digital content business by combining it with Metaverse.

ERC-20	ERC-165	ERC-233
ERC-621	ERC-721	ERC-777
ERC-827	ERC-884	ERC-1155

Table 2.
ERC(Ethereum request for comment)-number.

4.2 Ethereum-based Metaverse

The Metaverse is a three-dimensional virtual space where social and economic activities are commonly used just like the real world. NFT plays a role of mediating interaction and proving private property within the Metaverse world. An example of NFT application is the CryptoKitties. It is a blockchain-based cat reproduction game. CryptoKitties is an Ethereum ERC-721 token-based DApp [11]. Game users are given only one cat in the world in CryptoKitties. Cat digital assets have a rarity because they contain a separate unique recognition value, unlike existing virtual assets. In general online games, when the service is terminated, there is a problem that the character developed in the game can no longer be owned. However, digital assets with NFT technology can be distributed and stored by individual participants connected to the network to prove ownership.

Decentraland implemented the concept of real estate in the Metaverse by combining virtual reality and blockchain technology [12]. Decentraland made it possible to purchase land, a virtual real estate, using MANA, an ERC-20 token. Users can freely place buildings on land purchased from Decentraland, earn income by attaching billboards to buildings, or open exhibitions by collecting rare digital content. Land ownership and other collectible items are ERC-721 non-fungible tokens. These unique assets are made through Ethereum smart contracts and allow owners to prove ownership on the blockchain ledger. Cryptocurrency MANA can be purchased on exchanges and can also be used to purchase digital goods and services around the world.

Enjin Coin is a cryptocurrency project created for game item trading, and is an integrated platform for creating blockchain-based games. Enjin is a smart contract platform based on the Ethereum blockchain, and is a protocol and cryptocurrency that supports the crypto needed to create, manage, and implement virtual goods for game developers, content creators, and game communities. Ethereum-based ERC-20, ERC-721, and ERC-1155 token items can be stored and managed in a mobile cryptocurrency wallet. Enjin Coin guarantees the ownership and currency value of game items used in all games. When Enjincoin is applied as a currency in Metaverse, it can be used not only as a currency in Metaverse, but also in the real world with the value of currency.

5. Complementary point of Blockchain and artificial intelligence

5.1 Artificial intelligence and blockchain

Through the cognitive revolution, the agricultural revolution, and the scientific revolution, humans have entered the stage of connected intelligence, which uses the combined intelligence of humans and machines. As in the movie The Matrix, a symbiotic relationship between humans and machines has begun, and artificial intelligence and blockchain technology are accelerating this.

Artificial intelligence is reaching a stage where prediction and creation are possible through pattern recognition and learning using large amounts of data. And artificial intelligence is helping people to reduce repetitive tasks and human errors. Blockchain technology has deeply entered our society as a digital asset and is developing into a safe and reliable transaction through decentralization. Artificial intelligence is the core of the Forth industrial revolution, and it can be integrated with blockchain technology to make both artificial intelligence and blockchain more powerful [13]. Artificial intelligence and blockchain can change business models and have a transformative impact on society.

5.2 Blockchain for artificial intelligence

Artificial intelligence has a centralized nature where data is centrally managed and stored, making it a target for hacking and manipulation, which can lead to data tampering. In addition, since the source and reliability of the source for generating data are not guaranteed, there are many errors and risks. The blockchain capabilities of immutability, origin and control mechanisms have the potential to address the shortcomings of artificial intelligence and improve the accountability of trust, privacy issues and decisions. The combination of blockchain and artificial intelligence can help enable trusted digital analysis and decision-making on vast amounts of data. And it can be used to create secure data sharing and make artificial intelligence explainable, as well as regulating trust between devices that cannot trust each other [13].

5.3 Artificial intelligence for blockchain

Integrity of blockchain data is guaranteed. However, the security of applications built on top of the blockchain platform is not secure. Also, when a new block is added to the blockchain and consensus of all nodes is required, a problem arises that it cannot be used efficiently in fields that require high speed. When an error or vulnerability is found in the script of a smart contract and needs to be corrected, the irreversibility of the blockchain can hinder it. The case of hacking tens of millions of dollars in crypto currencies using vulnerabilities in smart contract algorithms reminds us of the need for agents that can immediately compensate for imperfect algorithms [14]. In such cases, machine learning systems of artificial intelligence can improve the security of blockchain applications, adjust dynamic parameters for scalability, and provide effective personalization and governance mechanisms.

Netflix provides a list of related movies related to your favorite movies, but this is the result of Netflix's central server analyzing personal information. If you do not provide personal information to Netflix, your personal information will be protected, but you will not be provided with personal taste analysis. Instead of collecting data on a central server, you might consider making use of data stored on a decentralized blockchain. However, in the case of a public blockchain, anyone can look into the transaction ledger, so there may be a risk of invasion of privacy as well. Although it is possible to allow individuals to directly control personal information in the blockchain, there is a risk of incurring a lot of cost. Artificial intelligence can provide customized services to individuals without violating personal information. Artificial intelligence can perform analysis on the user's local device and not perform analysis that is not permitted in advance. Artificial intelligence can realize decentralization so that real individuals have control over personal information [15, 16].

6. Blockchain and artificial intelligence encounter in the Metaverse

Blockchain plays an important role in implementing the economic system in Metaverse. The economy of Metaverse without blockchain will eventually be controlled by someone. If the blockchain is not supported, it is difficult for resources or goods used in the Metaverse world to be recognized for their value or to have economic interactions equivalent to the real economy. NFT-based blockchain technology further activated the Metaverse. With the advent of WEB 3.0 and Blockchain 3.0, Metaverse becomes the world to realize it.

In the Metaverse, people appear by scanning themselves in 3D or transforming them into avatar characters. Characters in the Metaverse are recognized as beings like clones in real life, not just game characters. In the Metaverse, besides their own avatars, they create things that can express their uniqueness. And to prove this, the NFT technology of the blockchain is used.

6.1 High quality learning data

In the real world, the problem of people's time, labor, and cost is easily replaced by using artificial intelligence in Metaverse. In the real world, when delivering news, you have to go through a lot of work, such as recruiting an announcer, shooting in a studio, and editing video. However, in the virtual space, by utilizing an artificial intelligence announcer, it is possible to deliver urgent and important news quickly and continuously for a long time. In order to deliver news in the Metaverse, it is necessary to learn the facial expressions, muscle movements, voices, nuances, and gestures of real announcers. When learning by receiving a long-time news video from a broadcaster to make an artificial intelligence announcer video, we extract only the part where the voice of another reporter and noise-free data, and the announcer's face and voice come out clearly toward the camera, and only detect a specific person techniques must be applied. If you use blockchain meta-information when searching for various data like this, you can select only the pure data necessary for learning and induce high-quality learning. Metadata stored within the blockchain block makes the necessary high-quality data selectively available. It is created as reliable data in the Metaverse, which increases the number of users who use the Metaverse.

6.2 Reusable data

Recently, creative activities in Metaverse are often developed using artificial intelligence instead of real people. When artificial intelligence artists creates works, they learn about the trends and styles of the works, and then express what they learned for creation. In the past, a lot of data was used for style analysis. Now, artificial intelligence artists store the data in the distributed ledger so that it can be easily selected and reused. Acquiring more data and practicing iteratively reduces the chance of selecting the wrong data and shortens validation time.

6.3 Stable decentralized network

Metaverse is a virtual 3D environment that requires a large amount of data and server capacity. However, controlling through a central server can incur a lot of cost. By utilizing the distributed environment system of blockchain, it is necessary to have a network system that can use the Metaverse environment with each individual's PC computing. When individuals control the Metaverse environment they want to use or view, the burden of centrally managing vast amounts of data can be reduced. It can also prevent some big tech companies from monopolizing the Metaverse environment.

6.4 Privacy

There is a need for a system that can govern so that ethical problems do not arise with respect to persons belonging to the Metaverse. Only the publicly available information about real and virtual people should be made known. And a personal information security system should be applied to prevent any damage to privacy. However, digital virtual people have no legal basis, so they are easier to manipulate or transform photos than real people, and there is a concern that the wrong algorithm may be applied, which may lead to serious racial and gender discrimination. With regard to personal information, it can be safely protected with blockchain to prevent external attacks. If personal information is erroneously altered, it can be managed responsibly with a clear path that can be traced based on the time of occurrence.

6.5 Distinguishing between virtual and real

In order to create a stable environment in which users are not confused in the Metaverse, a device that can distinguish between artificial intelligence and real people is needed. The fictional characters used in the Metaverse have now reached a level where it is difficult to distinguish the real from the fake from the human point of view. A reliable data construction system is needed to inform the comparison and judgment between real and fictional people. Data should be transparent and descriptive so that fake news and fake photos can be identified. Data content should be stored in a blockchain so that people can accurately know and understand the data generated by artificial intelligence and know the detailed history if desired. Blockchain technology can be used as a data to explain the data generated by artificial intelligence.

6.6 Rich content

We are using artificial intelligence technology as a way to imitate human behavior and replace it. Artificial intelligence analyzes the user's behavioral patterns such as words and messages in the Metaverse to predict the user's personality, intellectual level, and economic level. Metaverse uses artificial intelligence to create human-like voices and unique content. These data can be automatically converted into games, YouTube, news, advertisements, and lecture materials by simply inputting simple information. It is possible to create vast pattern content that imitates human behavior by using artificial intelligence technology with the vast data needed for the Metaverse world. With blockchain, personal information can be safely protected and various types of content can be created more abundantly.

6.7 Economic virtuous cycle

In investment and business, artificial intelligence can be used to make decisions about which data to use. It is important to have more reliable data in changing forecasts. If blockchain data is used, more reliability can be guaranteed through history management, thereby increasing the reliability of business predictions. In addition, the Metaverse Marketplace can be further activated through the payment of tokens and coins based on blockchain technology.

7. Conclusion

In the Metaverse, various and large amounts of secondary and tertiary data are generated due to the activities of many users. In the blockchain-based Metaverse, this data has a unique identification tag and is used as traceable data. Such data is becoming a good material for artificial intelligence in the Metaverse. Metaverse uses artificial intelligence and blockchain technology to create a digital virtual world where you can safely and freely engage in social and economic activities that transcend the limits of the real world, and the application of these latest technologies will be accelerated. Artificial intelligence and blockchain technology are expected to play an essential role in the ever-expanding world of the Metaverse.

Blockchain and AI Meet in the Metaverse
DOI: http://dx.doi.org/10.5772/intechopen.99114

Conflict of interest

The authors declare no conflict of interest.

Author details

Hyun-joo Jeon[1*], Ho-chang Youn[2], Sang-mi Ko[3] and Tae-heon Kim[4]

1 Konkuk University, Chungjoo, South Korea

2 SCCA, Seoul, South Korea

3 University of Michigan, Dearborn, USA

4 Chungnam National University, Daejeon, South Korea

*Address all correspondence to: jenny1603@naver.com

IntechOpen

References

[1] J. M. Smart, J. Cascio, and J. Paffendorf, Metaverse Roadmap Overview; Acceleration Studies Foundation; 2007.

[2] Oh Jeong-seok, Youn Ho-chang, A Study on New Concept of Culture; Proceedings of the Korea Contents Association Conference; 2004. p.54-60.

[3] Oh Jeong-seok, Youn Ho-chang, Jean Hyun-joo, Kim Tea-heon, A Study on a Creativity of SeaCircle; Proceedings of the Korea Contents Association Conference; 2013. p.61-62.

[4] Lee Min-hwa, Examples of Advanced Countries in the Forth Industrial Revolution and Korea's Response Strategy; Advanced Policy Series; Hansun Foundation; 2017. p.14-107.

[5] Gretchen Kell, Unforgotten: COVID-19 era grads to be celebrated virtually this Saturday, Media Relations, Berkely News [Internet]. 2020. Available from: https://news.berkeley.edu/2020/05/14/unforgotten-covid-19-era-grads-to-be-celebrated-virtually-this-saturday [Accessed:2021-01-30]

[6] Roblox Studio, Anything you can imagine, make it now!, Roblox [Internet]. 2021. Available from: https://www.roblox.com/create [Accessed:2021-01-30]

[7] The Sandbox [Internet] 2020. Available from: https://installers.sandbox.game/The_Sandbox_Whitepaper_2020.pdf [Accessed:2021-01-30]

[8] Park Hyun-young, The Combination of Metaverse and Blockchain, Digital Daily [Internet]. 2021. Available from: http://m.ddaily.co.kr/m/m_article/?no=208850 [Accessed: 2021-03-30]

[9] Melanie Swan, Blockchain: Blueprint for a New Economy, O'Reilly Media, 2015.

[10] Lee Je-young, Blockchain 3.0 Era and Future of Cryptocurrency; Future Info Graphics; Research Center for New Industry Strategy; 2021.

[11] Park Kyung-ho, ERC-number, Television Creative Contents, 2021. Available from: https://www.tvcc.kr/article/view/10277 [Accessed: 2021-04-30]

[12] Decentraland, Metaverse Property [Internet] 2021. Available from: https://metaverse.properties/buy-in-decentraland [Accessed:2021-03-30]

[13] Bhaskar Chavali, Sunil Kumar Khatri and Syed Akhter Hossain, AI and Blockchain Integration; In: Proceedings of International Conference on Reliabilit; Infocom Technologies and Optimization; 2020.

[14] Lucas Mearian, Blockchain has Five Problems, IT World [Internet] 2017. Available from: https://www.itworld.co.kr/news/107168 [Accessed: 2021-03-30]

[15] Thang N. Dinh, My T. Thai : AI and Blockchain: A disruptive integration. Computer: 2018. p. 48-53. DOI: 10.1109/MC.2018.3620971

[16] Kim Yun-kyung, Changes in the content production environment as seen by AI announcers; Reimaging the Future, TechM Conference, 2021.